GENETICS – RESEARCH AND ISSUES

# A COMPREHENSIVE GUIDE
# TO GENETIC COUNSELING

# GENETICS – RESEARCH AND ISSUES

Additional books and e-books in this series can be found on Nova's website under the Series tab.

# A COMPREHENSIVE GUIDE TO GENETIC COUNSELING

## BENJAMIN A. KEPERT
### EDITOR

nova
science publishers
New York

Copyright © 2020 by Nova Science Publishers, Inc.

## NOTICE TO THE READER

## Library of Congress Cataloging-in-Publication Data

Names: Kepert, Benjamin A., editor.
Title: A comprehensive guide to genetic counseling / Benjamin A. Kepert, editor.
Identifiers: LCCN 2019056880 (print) | LCCN 2019056881 (ebook) | ISBN
   9781536169751 (paperback) | ISBN 9781536169768 (adobe pdf)
Subjects: LCSH: Genetic counseling. | Genetic screening.
Classification: LCC RB155.7 .C66 2019 (print) | LCC RB155.7 (ebook) | DDC 362.196/042--dc23
LC record available at https://lccn.loc.gov/2019056880
LC ebook record available at https://lccn.loc.gov/2019056881

*Published by Nova Science Publishers, Inc. † New York*

# CONTENTS

# PREFACE

In this compilation, the authors describe the psychological consequences and effects of predictive cancer genetic testing in both breast and ovarian cancer. Predictive genetic tests are being offered to an increasing number of women as the availability and public awareness of genetic testing increases.

Next, A Comprehensive Guide to Genetic Counseling assesses the parental psychosocial implications, such as emotions and coping, regarding the earlier diagnosis of Usher syndrome via genetic testing compared to parents of children who were diagnosed later via ophthalmologic findings.

The closing chapter reports on a comprehensive analysis of the studies performed on genetic counseling, particularly those pertaining to the Israeli Arab society. Furthermore, an overview of the various socio-demographic, economic, cultural and religious aspects is provided.

Chapter 1 - The aim of this chapter is to describe the psychological consequences and effects of predictive cancer genetic testing in both breast and ovarian cancer. Predictive genetic tests are being offered to progressively more women as the availability and public awareness of genetic testing increases. Usually these tests are offered within a counselling protocol and research suggests that women will have differing emotional responses to genetic testing and its varying results. The authors want to outline the impact of undergoing these tests, the emotional state of women and their mental representations during the genetic counseling process. The authors also describe both the preventive and supportive role of genetic counseling across the various phases from pre-test to post-test. Often, cancer genetic counseling situations are complex, and the involvement of the patients is just as important as that of the multidisciplinary team working with them. Counselling must help patients and their caregivers in healthcare settings make an informed decision regarding genetic testing and risk management and share worries, fears, uncertainties and expectations to try to experience the testing process with the least possible distress. The authors' hope is that they provide a more precise picture of what surrounds cancer genetic testing and genetic counseling in order to inform both patients and healthcare providers of the importance of not only the physical and medical aspects, but also of the psychological and social aspects of people undergoing this process. The authors will also discuss the recent changes regarding access to genetic testing and future research areas regarding gynaecological cancer genetic testing, prompted by the development of treatment-focused genetic testing (TFGT) for breast and ovarian cancer.

Chapter 2 - *Purpose*: The aim of this study was to assess the parental psychosocial implications, such as emotions and coping, of earlier diagnosis of Usher syndrome via genetic testing compared to parents of children who were diagnosed later via ophthalmologic findings.

*Method*: Thirty-six participants were recruited through an online posting on the Usher Syndrome Coalition website. Two comparison groups were formed based on the method of diagnosis (i.e., genetic diagnosis vs. ophthalmologic diagnosis). Semi-structured interviews were recorded and transcribed. Comparison, using thematic and statistical analysis, of psychosocial impact on parents of children diagnosed early (via genetic testing) and later (based on ophthalmologic findings) was completed.

*Results*: There were no statistically significant differences in emotions between the two groups of participants, suggesting that earlier diagnosis via genetic testing does not lead to increased anxiety or psychosocial issues for parents. Additional themes identified from parent interviews and their application to patient care are described.

*Conclusion*: Earlier diagnosis of Usher syndrome via genetic testing does not cause a more harmful emotional impact than later diagnosis via ophthalmologic findings. In fact, there are multiple benefits to earlier diagnosis via genetic testing. Earlier diagnosis allows parents to emotionally process and prepare the child for independence throughout life.

Chapter 3 - Genetics is considered to be a scientific study of the mechanism of inheritance and causes of variation in all of the living organisms related by descent. Although man has always been aware that different individuals vary between themselves but children do resemble their parents in some way. The scientific basis for these observations was given thrust during the past 150 years. The clinical application of this knowledge has become widespread with most progress made in the past few decades. Genetic diseases were known long ago, but they were poorly understood. A more specific measure of the impact of genetic disorders is their role in mortality and morbidity. In an attempt to cope with these genetic diseases, the process of genetic counseling (GC) was newly introduced. Thus, GC could be defined as the process of advising individuals and families affected by or at risk of genetic disorders to help them understand and adapt to the medical, psychological and familial implications of genetic contributions to disease.

The advent of prenatal diagnosis and the ability to detect an inborn error of metabolism, chromosome disorder, or congenital malformation in utero, early enough to allow termination of pregnancy, has added a new dimension to genetic counseling.

Almost, all Arab countries are characterized by high rates of consanguineous marriages with a common founder effect and do have common and rare gene-mutations that give rise to genetic disorders. As the Arab society of Israel has unique socio-demographic and cultural characteristics with some common and rare genetic disorders and high rates of consanguineous marriages, that differ from the rest of the population, therefore, genetic counseling for this population should be carefully designed.

The authors report on a comprehensive analysis of the local studies performed so far on genetic counseling particularly those pertaining to the Israeli Arab society. This review is an attempt to locate certain procedures that should be implemented while extending the genetic counseling services to all targeted population. Furthermore, an overview of the various socio-demographic, economic, cultural and religious aspects will be taken into account.

In: A Comprehensive Guide …          ISBN: 978-1-53616-975-1
Editor: Benjamin A. Kepert          © 2020 Nova Science Publishers, Inc.

*Chapter 1*

# PSYCHOLOGICAL ASPECTS IN BREAST AND OVARIAN CANCER GENETIC TESTING

*Valentina Elisabetta Di Mattei[1,2],*

*Martina Mazzetti[2] and Martina Bernardi[3]\**

[1] Faculty of Psychology, Vita-Salute San Raffaele University,
Milan, Italy
[2] Clinical and Health Psychology Unit, Department of
Clinical Neurosciences, San Raffaele Hospital, Milan, Italy
[3] Languages Department, Univeristy of Parma, Parma, Italy

## ABSTRACT

The aim of this chapter is to describe the psychological consequences and effects of predictive cancer genetic testing in both breast and ovarian cancer. Predictive genetic tests are being offered to progressively more women as the availability and public awareness of genetic testing increases. Usually these tests are offered within a counselling protocol and research suggests that women will have differing emotional responses to genetic testing and its varying results. We want to outline the

\*Corresponding Author's E-mail: martina.bernardi@unipr.it.

impact of undergoing these tests, the emotional state of women and their mental representations during the genetic counseling process. We also describe both the preventive and supportive role of genetic counseling across the various phases from pre-test to post-test. Often, cancer genetic counseling situations are complex, and the involvement of the patients is just as important as that of the multidisciplinary team working with them. Counselling must help patients and their caregivers in healthcare settings make an informed decision regarding genetic testing and risk management and share worries, fears, uncertainties and expectations to try to experience the testing process with the least possible distress. Our hope is that we provide a more precise picture of what surrounds cancer genetic testing and genetic counseling in order to inform both patients and healthcare providers of the importance of not only the physical and medical aspects, but also of the psychological and social aspects of people undergoing this process. We will also discuss the recent changes regarding access to genetic testing and future research areas regarding gynaecological cancer genetic testing, prompted by the development of treatment-focused genetic testing (TFGT) for breast and ovarian cancer.

## MEDICAL INTRODUCTION TO HEREDITARY BREAST AND OVARIAN CANCERS

Hereditary gynaecological cancer syndromes arise from a germline mutation, inherited from one's parents, with the result of an increased risk of cancer development relative to that of the general population without the mutation in a cancer susceptibility gene. Specifically, a germline mutation in *BRCA1* or *BRCA2* (on chromosomes 17 and 13 respectively) results in a significantly elevated lifetime risk of developing breast and ovarian cancer. These two mutations account for the majority (over 90%) of hereditary breast and ovarian cancer (hereafter HBOC) (Arindem and Soumen 2014; King, Marks, and Mandell 2003). Genetic susceptibility to breast or ovarian cancer might also be associated with mutations in other genes, some of which are associated with known hereditary cancer syndromes. Some of these genes include: *p53*, *PTEN*, *CDH1*, *STK11*, *MLH1*, *MSH2*, *MSH6* and *PMS2* (Paluch-Shimon et al. 2016). A mutation in one of these genes means that there is an incapacity to regulate cell death and there is uncontrolled cell growth, which both lead to cancer.

The estimated prevalence of *BRCA1* and *BRCA2* mutations is dependent on the population and can vary between 1/300 to 1/800, respectively. In some populations there are mutations that are particularly prevalent, for instance, in the Askenazi Jew population, 2.5% of people will harbour the BRCA1 mutation (Roa et al., 1996). The prognoses for these conditions depend on the stage at which the cancer is diagnosed (Petrucelli, Daly & Feldman, 2013).

The risk of developing HBOC is suspected on the basis of a family history and diagnosis is made following a molecular genetic test, usually carried out via a blood sample. Genetic testing is not usually offered to unaffected individuals, unless a mutation has been identified in the family.

A genetic test is usually carried out when one of the following criteria are met (Mayo Foundation for Medical Education and Research, 2019):

1) A history of breast cancer diagnosed at a young age (<50 years of age);
2) A personal history of triple negative breast cancer diagnosed at age 60 or younger;
3) A personal history of breast cancer affecting both breasts (bilateral breast cancer);
4) A personal history of both breast and ovarian cancers;
5) A personal history of ovarian cancer;
6) A personal history of breast cancer and one or more relatives with breast cancer diagnosed at age 50 or less, one relative with ovarian cancer, or two or more relatives with breast or pancreatic cancer;
7) A history of breast cancer at a young age in two or more close relatives, such as parents, siblings or children;
8) A male relative with breast cancer;
9) A family member who has both breast and ovarian cancers;
10) A family member with bilateral breast cancer;
11) A relative with ovarian cancer;
12) A relative with a known BRCA1 or BRCA2 mutation;
13) Ashkenazi Jewish ancestry, with a close relative who has breast, ovarian or pancreatic cancer at any age.

Risk reduction can include lifestyle modifications. Various studies have found that breastfeeding can reduce the risk of breast cancer in BRCA1/2 carriers (Kotsopolous et al. 2012; Jernström et al. 2004). Exercising regularly, maintaining a healthy body weight and limiting alcohol consumption should be advised, as well as avoidance of hormone replacement therapy (Paluch-Shimon 2016). Screening is another way of reducing the risk of developing HBOC. Carriers are advised to be "breast-aware" and seek medical attention immediately if they feel any changes in their breasts. For the high-risk population, annual breast Magnetic Resonance Imaging (MRI) (or breast ultrasonography if MRI is unavailable) from age 25 is recommended in addition to annual mammography from age 30 (Paluch-Shimon 2016). Adherence to screening regimens usually exceeds 50%, and reaches over 75% in women aged 40 years or more in the USA (Buchanan et al., 2017).

With regards to risk-reducing surgery, bilateral mastectomy seems to be a very effective method for reducing breast cancer in carriers of BRCA1/2 (Domchek et al. 2010; Evans et al. 2009). Bilateral risk–reducing mastectomy (BRRM) lowers breast cancer risk by about 90% in asymptomatic individuals (Lostumbo et al. 2010). BRRM may involve complete removal of the breasts, including the nipples (total mastectomy), or it may involve removal of as much breast tissue as possible while leaving the nipples intact (subcutaneous or nipple-sparing mastectomy). Subcutaneous mastectomies preserve the nipple and allow for more natural-looking breasts if a woman decides to have breast reconstruction surgery afterwards. It must be noted that total mastectomy provides the greatest breast cancer risk reduction since more breast tissue is removed in this process than in a subcutaneous mastectomy (Guillem et al. 2006). Advantages, disadvantages, risks of complications and psychosocial impact should be discussed with the individual carrier. Contralateral risk-reducing mastectomy (CRRM) among patients with a previous breast cancer diagnosis can also be considered (Heemskerk-Gerritsen et al. 2015). Some women diagnosed with cancer in one breast, particularly those who are known to be at very high risk, may consider having the other breast (contralateral breast) also removed, even though there is no evidence of cancer in that breast. Prophylactic surgery to remove a contralateral breast during breast cancer surgery lowers the risk of breast cancer in that breast (King et al. 2011; Lotsumbo et al. 2010). Immediate or delayed prophylactic CRRM in women who may be at risk of developing breast cancer has been documented to reduce the risk of contralateral cancer up to 95% (Lostumbo et al. 2010).

Risk-reducing salpingo-oophorectomy (RRSO), which involves the removal of the ovaries and fallopian tubes, represents the most effective measure for ovarian cancer prevention; in particular, this procedure has consistently been shown to reduce cancer risk by 80%–90% (Finch et al. 2014; Marchetti et al. 2014) and results in a 77% reduction in overall mortality (Finch et al. 2014). RRSO can be carried out on its own or together with bilateral prophylactic mastectomy in premenopausal women who are at very high risk of breast cancer. Removing the ovaries in premenopausal women reduces the amount of oestrogen that is produced by the body. As oestrogen promotes the growth of some breast cancers, reducing the amount of this hormone in the body by removing the ovaries may slow the growth of those breast cancers (Guillem et al. 2006). RRSO has repeatedly been reported in several retrospective and prospective studies to reduce the risk of breast cancer among *BRCA1/2* mutation carriers when carried out in premenopausal women (Rebbeck et al. 2009). However, a recent prospective cohort study, controlling for potential biases, suggested that no benefit in breast cancer risk reduction existed following RRSO (Heemskerk-Gerritsen et al. 2015).

Currently, no agent has been proven by interventional trials to possess chemopreventive properties against ovarian cancer in high risk BRCA1/BRCA2 mutation positive women. According to Kathawala et al. (2018), more studies and clinical trials are needed to explore the opportunities that chemoprevention may bring in ovarian cancer.

However, in breast cancer the situation is different. The Breast Cancer Risk Assessment Tool by the National Cancer Institute and the National Surgical Adjuvant Breast and Bowel Project Biostatistics Center can be used to assess the eligibility of women for chemoprevention. Women most likely to benefit from preventive therapy include those at high risk below

50 years of age and those with atypical hyperplasia. Pruthi et al. (2015) noted that certain physician and client barriers limit the acceptance and adherence of chemopreventive strategies and generally uptake of chemoprevention has been low. Nevertheless, tamoxifen and raloxifene, which are selective oestrogen receptor modulators, as well as two aromatase inhibitors, exemestane and anastrozole, have been demonstrated to reduce breast cancer incidence significantly in randomized controlled trials in women at increased risk of developing the disease (Cuzick et al. 2014; Fisher et al. 1998; Fisher et al. 2005; Goss et al. 2011; Vogel et al. 2006; Vogel et al. 2010). In a breast cancer prevention trial tamoxifen reduced the risk of breast cancer in both pre- and postmenopausal women at increased risk of the disease by approximately one-half (Fisher et al. 1998). Raloxifene retained 76% of the effectiveness of tamoxifen in preventing invasive disease and has lower risks of developing uterine cancer or venous thromboembolic episodes (Vogel et al., 2010).

Exemestane has been shown to reduce breast cancer risk by 65% (Goss et al. 2011) in high risk postmenopausal women while anastrozole showed a 53% reduction in breast cancer risk (Cuzick et al. 2014). As these preventive treatments can all have serious side effects and physicians have less time to counsel women about these options, adherence to prophylactic chemopreventive regimens seems to remain low. Communication regarding chemoprevention options should be included in the decision-making segment of the medical consultation. The use of these chemopreventive agents could significantly reduce the incidence of oestrogen receptor-positive breast cancer but will have no impact on oestrogen receptor-negative breast cancer (Pruthi et al. 2015). Current research is, therefore, focused on pinpointing alternative mechanisms by which biologically active compounds can lower the risk of all breast cancer subtypes including oestrogen receptor-negative ones. Promising agents are currently being developed and include inhibitors of the ErbB family receptors, COX-2 inhibitors, metformin, retinoids, statins, poly-(ADP-ribose) polymerase inhibitors, and natural compounds (Beate & Brown, 2014).

Deciding whether to undergo surgical or chemopreventive interventions is a very personal and difficult decision and genetic counsellors, oncologists and healthcare workers can help the individual to understand all the risks, limitations and benefits involved. Surgical procedures have a dramatic psychological and social impact, effect on fertility, and are not always completely effective in preventing hereditary cancer. However, studies examining psychosocial aspects of BRRM and RRSO have, for the most part, demonstrated a favourable impact among women undergoing the procedure both in short- and long-term follow-up (Collins et al. 2018; Shigehiro et al. 2015). When it comes to chemoprevention it seems that many women are not willing to accept the side effects of a preventive therapy for many years. To overcome this obstacle education of the public and medical community with evidence-based risk and benefit information is needed. Perhaps alternating dose schedules should be investigated to reduce frequent and rare serious side effects (Beate and Brown, 2014).

## HBOC GENETIC TESTING AND COUNSELLING

A growing body of evidence supports the benefits of genetic risk assessment, genetic testing, and the efficacy of clinical management in those with certain hereditary cancer syndromes.

Genetic counseling and risk assessment is the process of identifying and counselling individuals at increased risk of developing cancer, and distinguishing between those at high risk, those at a modestly increased risk, and those at average risk (Bronson et al. 2012).

Usually the genetic counseling and testing procedure takes the following form:

1) The *selection* process for a woman to take part in the genetic testing procedure often begins with a general practitioner's, oncologist's or gynaecologist's referral. These experts may feel there is a necessity for the counselee to undergo a genetic test

because there are signs that they may have a hereditary form of cancer. On the other hand, some counselees feel that they want to undertake the test themselves. Whether or not to subject the client to the test is evaluated by a multidisciplinary team for every individual; it is not always granted, especially if the team feels that the patient cannot cope with the test results.

2) If the multidisciplinary team agrees that the testing can be undertaken, next during the *pre-testing stage*, the subject learns about the hereditary, familial and sporadic forms of cancer and the methods available to identify the risk of developing these. The pedigree construction (family medical history) for at least three or four generations allows the geneticist to estimate the counselee's risk profile; by examining an individual's clinical history, the geneticist takes into account all the risk factors and family history. Physical examinations may also be carried out at this stage (Bronson et al., 2012). The pre-test psychological interview aims to identify users who experience cancer genetic counseling as more stressful, assessing their coping strategies and psychological distress, and programming a possible intervention of personalized psychological support. Understandably, there may be emotional reactions to the pre-test phase including: anxiety, fear, and shock at having to answer all the personal and family history questions and from all the information the genetic counsellor may divulge. Assessing the psychosocial impact in the pre-test phase often provides clues about how the counselee and her family may understand and cope with disclosure of genetic testing information (Edwards et al. 2008; Pieterse et al. 2007).

3) In case of suspected inherited risk, a genetic test is offered (*testing stage*) when the following conditions apply: an individual has a personal or family history suggestive of a HBOC risk (see previous section), the genetic test can be adequately interpreted (ASCO 2003), testing will influence medical management of the client and her relatives, the benefits outweigh the potential risks, testing is voluntary and, the individual seeking testing (or their legal proxy)

can provide informed consent (ASCO 2003, Bronson et al. 2012). Users then have to wait a few months for blood test results. The genes are separated from the rest of the DNA and scanned for abnormalities.

4) Once the molecular analysis is complete, the interdisciplinary team communicates the test results to the counsellee (*post-test stage*): here the multidisciplinary team helps the individual read and understand the test result and also comprehend the implications this may have on them and their family. This should be a personalized interpretation of results, regardless of the test outcome and should be done in person when possible.

The information gathered is then used to develop a management plan for cancer screening, prevention, reproductive options and risk-reduction. Furthermore, notification of any at-risk family members will follow. Genetic counseling also includes client education about HBOC and assistance coping with the psychological responses that can occur in families at increased cancer risk (Trepanier et al. 2004).

## Treatment-Focused Genetic Testing (TFGT)

The diffusion of cancer genetic testing has become increasingly widespread in recent years, as it has been gradually acknowledged how the detection of BRCA1/2 mutations enables not only to predict breast and ovarian cancer risk, but also provides crucial information regarding the clinical course of the disease and may significantly influence decision-making about treatment options. The transition from the main predictive function of genetic testing to the comprehension of its important therapeutic value has been prompted by the development of new treatment agents, along with technological improvements in gene sequencing and decreasing costs associated with genetic testing (George, Kaye, and Banerjee 2017; Trainer et al. 2010;).

The increasingly strong proof of efficacy of poly-(ADP-ribose) polymerase (PARP) inhibitors in the treatment of BRCA-mutated ovarian cancer patients has represented a turning point in clinical practice, revealing the importance in identifying in advance the potential candidates to this biomarker-directed therapy. Specifically, in December 2014 the European Medicines Agency approved Olaparib as a maintenance therapy for patients with platinum-sensitive, relapsed, BRCA-mutated (germline or somatic), high-grade serous ovarian cancer. At the same time, the Food and Drug Administration licensed the implementation of this targeted agent for the treatment of patients with recurrent, germline BRCA-mutated, advanced-stage ovarian cancer who have already received three or more prior lines of chemotherapy (The US department of Health and Human Services 2014). Since the implementation of PARP inhibitors (pharmacological inhibitors of the enzyme poly-ADP-ribose polymerase) in ovarian cancer routine care, a growing need has arisen in knowing the BRCA-mutation status of each patient before treatment, in order to give them the opportunity to access a personalised clinical management.

Genetic mutational analysis plays a key role also in breast cancer treatment, as clinical trials have shown that BRCA status might inform the choice of specific neo-adjuvant chemotherapy regimens, in particular with respect to the use of platinum-based agents (Telli et al. 2015). Moreover, the detection of BRCA pathogenic variants may significantly shape patients' surgical management, guiding decisions regarding the extent of the surgery and the possibility of immediately performing a contralateral prophylactic mastectomy, as well as the evaluation of the appropriateness of radiotherapy (Weitzel et al. 2003).

These findings have prompted relevant changes in accessibility conditions to cancer BRCA1 and 2 genetic counseling and testing, which has traditionally been limited to individuals reaching a defined clinical threshold or satisfying a precise set of criteria (Eccles et al. 2016). In particular, the presence of a history of multiple familial cancer has been usually considered the main standard for genetic testing; however, this has resulted in a significant underestimation of BRCA 1 and 2 mutated-patients, as a substantial proportion of carriers do not actually present family history (Alsop et al. 2012; Møller et al. 2007; Trainer et al. 2010, Zhang et al. 2011). These limitations have led to the development of expanded selection criteria, which are actually centred on the evaluation of the histological subtype of the disease, irrespectively of familial and/or personal history of cancer (Demsky et al. 2013; George et al. 2016; Zhang et al. 2011), thus determining a significant increase in genetic testing request.

In order to address the need to provide rapid and extended access to genetic services, new models of genetic testing have been proposed. These new approaches - variously defined as "treatment-focused genetic testing" (TFGT) and "rapid genetic counseling and testing" (RGCT) - are based on the integration of BRCA mutational analysis as an important and routine part of the diagnostic process, with the objective of developing a streamlined pathway care which enables breast and ovarian cancer patients to receive individualised treatment (House of Lords 2009).

The first TFGT approach was developed within the Mainstreaming Cancer Genetic Programme at the Royal Marsden Hospital of London and represents a reference model for the incorporation of genetic counseling and testing in cancer care. Specifically, this model establishes that all women with non-mucinous ovarian cancer, independently of their age at diagnosis, should be offered the opportunity to undergo BRCA genetic testing in their routine clinical visits by one of the members of the oncology team (including nurses), provided the previous completion of a specific online training course. During the appointment, patients should be given an information sheet which clarifies the BRCA testing procedure and its possible implications. Mutational analysis results, which should be obtained within 3-4 weeks, will be sent directly to patients, in writing, along with information regarding its specific clinical meaning. Those who test positive should also receive an appointment with a specialised geneticist, in order to gain a better understanding of their personal cancer risk and discuss the available surveillance measures and risk-reducing strategies, as well as the potential implications for family members. In parallel, the oncologists should consider and elucidate how the genetic test result might influence decision-making about cancer patient treatment (George, Kaye, and Banerjee 2017).

The preliminary results regarding the effectiveness of this model have revealed that it seems suitable both for patients and health-providers (George et al. 2016), thus having fostered its increasing implementation in cancer centres worldwide. Furthermore, broad systematic studies, mainly conducted in research settings, such as the Genetic Testing in Epithelial Ovarian Cancer (GTEOC; Plaskocinska et al. 2016) and the DNA-BONus study (Høberg-Vetti et al. 2016), have confirmed the relatively good degree of acceptability of TFGT, underlining that patients tend to appreciate the clinical importance of timely testing (Gleeson et al. 2013; Meiser et al. 2012; Wevers et al. 2017). Moreover, research results suggest that the majority of patients do not consider TFGT as a significant stressor, nor report significant adverse effects on psychological well-being (Hoberg-Vetti et al. 2019; Weavers et al. 2017). Nonetheless, studies have recommended that clinicians should adopt particular caution and sensitiveness in proposing and implementing this approach (Augestad et al. 2017; Shipman et al. 2017).

Moreover, research conducted in clinical settings, despite being somewhat limited, seems to support this evidence, highlighting the feasibility and perceived appropriateness of TFGT, which has shown to significantly contribute to shared decisional process in cancer care and adequately respond to patients' and health professionals' expectations and needs (Schwartz et al. 2006; Weavers et al. 2012; Zilliacus et al. 2012). A recent qualitative research by Wright and colleagues (2018) has investigated in more detail the subjective experience of breast and ovarian cancer patients undergoing TFGT, examining reactions to initial test proposal and the reported motivations to the mutational analysis. Authors have found that interviewed patients were globally receptive to the integration of genetic testing at an early stage of their care, irrespective of cancer family history, cancer type and specific treatment pathway. At the same time, results have revealed how the majority of patients did not show a clear acknowledgment of the specific therapeutic value of TFGT, but tended to conceptualize it in terms of a merely predictive and diagnostic tool, thus identifying familial protection as one of the main reasons for undergoing it. This evidence suggests that clinicians should make efforts to clearly explicit the distinctive characteristics of TFGT and emphasize its relevance for individualised treatment, in order to enhance patients' personal involvement in this procedure.

## ACCESS TO CANCER GENETIC COUNSELING AND DECISION-MAKING ABOUT TESTING

Clinical availability of genetic testing for cancer predisposition genes is generating a major worldwide challenge to health care systems to provide suitable and accurate genetic services to underserved populations.

The choice to attend a cancer genetic service usually involves a complex decision-making process, which often begins months or even years before the actual access to genetic counseling. People who become aware of the possibility of having a genetic mutation but do not feel confident enough to test for it immediately have to face conflicting emotions about the way to manage this knowledge, as well as the perception of themselves and their future. The decisional process leads individuals to carefully weigh relative benefits and potential risks of genetic testing and compels them to negotiate both intrapersonal conflicts - related to deep-rooted feelings about one's identity - and interpersonal dynamics. Clearly, balancing the pros and cons and integrating complex and competing emotions represent burdensome and time-consuming tasks, which could also significantly affect decisions about when to undergo genetic testing.

Research results (Kinney et al. 2006; Scherr et al. 2014; Willis et al. 2017) indicate that a number of socio-demographic, psychosocial, clinical and logistical factors may influence participation in genetic testing and decisions among individuals who are at increased risk of carrying a HBOC susceptibility gene mutation. Specifically, results regarding the role of socio-demographic factors as predictors of the access to cancer genetic counseling are almost inconclusive: while socio-economic status, education level and marital status seem to have a significant impact on genetic testing uptake, no consistent evidence has been reported with respect to age, sex, ethnicity and parenthood (Willis et al. 2017). Among psychosocial variables, disease-specific distress and perceived breast or ovarian cancer risk have been described as the main factors influencing genetic counseling uptake. Other important predictors are represented by clinical factors, including the presence of a personal or familial history of cancer and the objective estimated risk (Meiser et al. 2005; Hirschberg et al. 2015).

Despite the benefits associated with cancer genetic counseling and testing, about 33% of eligible subjects seem to decline the participation or choose to delay their appointment (Keogh et al. 2009; Ropka et al. 2006). Studies which have explored factors implicated in the refusal of genetic testing indicate that cost and logistical issues, for instance concerns about financial insurance, potential employment discrimination and problems related to time and travel required to attend an appointment, represent important barriers to the access to genetic services (Backes et al. 2011; Bleiker et al. 2005; Crook et al. 2015; Dekker et al. 2013; Geer et al. 2001; Godard et al. 2007; Wakefield et al. 2011). Psychological issues could play a significant role, too: specifically, emotional concerns regarding the psychological impact of genetic counseling and the presence of high levels of emotional distress, especially in terms of depressive symptoms, seem to be associated with the refusal or the delay of genetic counseling and testing (Backes et al. 2011; Bleiker et al. 2005; Godard et al. 2007; Schlich-Bakker et al. 2007; Yoon et al. 2011). Moreover, people who feel unable to tolerate a state of uncertainty and who have difficulties in coping with the knowledge of an increased risk or with personal and clinical consequences of a positive test result might be reluctant to undergo genetic testing (Caruso et al. 2008; Geer et al. 2001). Indeed, one should consider that, especially in the oncological field, knowledge about the presence of genetic mutations can often be confusing and uncertain in meaning, thus not always offering clear and unequivocal advantages. Low perceived personal relevance of genetic testing, which is often related to the pre-assumption of being mutation-negative, represents another issue frequently endorsed as a reason of non-attendance of appointments and testing (Bleiker et al. 2005; Caruso et al. 2008; Culver et al. 2001; Godard et al. 2007; Schlich-Bakker et al. 2007; Yoon et al. 2011).

Moreover, the fact that cancer genetic testing has important family implications could make the decisional process even more complex, as individuals are compelled to balance their own informational and emotional needs with those of their family, thus negotiating autonomy and altruism. Studies confirm that families could significantly influence testing decisions: while in some cases family members encourage or press each other to undergo genetic analysis, in other situations relatives' obstacle or even block cancer genetic counseling attendance. Family opposition could be motivated by personal fears and anxieties about mutation inheritance, as well as by concerns regarding the communication of a positive test result and its potential deleterious impact on family relationships; the perception of the proband as emotionally frail or too young to handle the risk knowledge could further lead family members to hinder genetic testing uptake (Klitzman 2012; Willis et al. 2017).

Further, health professionals seem to play a key role in the decisional processes: clinicians' level of knowledge and confidence about cancer genetic testing and their personal attitudes and behaviours could deeply influence individual choices by encouraging genetic counseling attendance or, on the contrary, by hampering it, more or less explicitly. In particular, research results highlight that sub-optimal referral practices and difficulties in doctor-patient communication, frequently due to a lack of adequate training in genetic counseling, could significantly deter genetic testing uptake (Petzel et al. 2013).

These issues underline the necessity to establish effective programs to facilitate information circulation and train healthcare care providers to promote communication with clients about cancer risk and cancer genetics. Furthermore, ulterior research is needed on the delivery of community-based primary care and specialist provider referral to cancer genetic services. Thus, access to genetic services can be increased among people from differing sociodemographic backgrounds. As clinical genetic testing becomes more diffused and the use of genetic information becomes more effective, there is a need to comprehend how to communicate genetic information adequately and supply genetic services for adult onset hereditary cancers to diverse populations. A more in-depth knowledge of factors influencing testing participation such as access problems, education levels and cultural factors such as the role of religious beliefs, in medical decision-making will be imperative if we are to provide cancer genetic services that are responsive to all populations. Disseminating information about the availability of testing and cancer genetics as well as development of culturally sensitive approaches is crucial for this to happen (Kinney et al. 2006).

## PSYCHOLOGICAL IMPACT OF TEST RESULTS

Cancer genetic test results can be classified into the following categories: a) *positive results*, which indicate the presence of pathogenic mutations, *id est,* alterations in the gene sequence associated with a significant increase in the likelihood of developing cancer; b) *true negative results*, indicative of the absence of a detectable mutation, which is already known in the counselee's family; c) *inconclusive or uninformative results*, which refer to situations in which no known mutation is detected, despite the presence of a strong history of breast/ovarian cancer in the family; d) *variants of uncertain clinical significance* (VUS), that is, unclassifiable variants in the assayed gene, for which too little scientific data are currently available to interpret them as benign polymorphisms (with no significant association with disease risk) or pathogenic mutations. These variants are routinely reclassified by laboratories on the basis of new collected evidence and, when this occurs, updated reports are sent to healthcare providers, who, in turn, should disclose the information to clients (Hoffman-Andrews 2018).

Emotional reactions and psychological adjustment to test results can vary widely over time, from the immediate post-test phase until years afterwards, representing a dynamic process, whose development and final destiny depend not only on the test result itself but also on a range of complex intrapersonal and relational factors. Indeed, the fact that genetic tests provide information about one's future, even if only in terms of probability, induce individuals to think deeply about their meaning and the way they can integrate this knowledge into their identity and into the personal narratives of their lives. In this sense, Klitzman (2012) suggests that a genetic test result could be seen as a sort of Rorschach test, whose interpretation can vary in function of a complex system of cognitive, emotional, interpersonal and socio-cultural issues, needs and beliefs.

## Positive Test Result

Understandably, a positive test result can produce adverse psychological reactions, especially in the immediate post-test phase, leading individuals to feel stunned and disappointed by the discovery of being carriers of a pathogenic mutation. Shock could arise from the fact that the communication of this information makes the notion of risk more actual and concrete than in the pre-test phase and is frequently considered as representative of a real and inevitable prospective of illness (Di Mattei et al. 2018). Indeed, even those who had assumed to carry a pathogenic mutation often feel unprepared to handle this knowledge on an emotional level. Shock and confusion could also be due to the amount and the inherent complexity of information to which counselees are exposed to during the communication of the test result, as well as to the difficult decisions about risk management (i.e., surveillance versus prophylactic surgery) they are confronted with.

Moreover, it is important to consider that, as genes, to some degree, shape who we are and our future, the genetic test could bring out existential questions about one's self-perception and sense of identity. Individuals who receive a positive test result are thus compelled to find a personal way to integrate the knowledge about their risk status into their self-concept and biographies, in an attempt to preserve a coherent narrative. The extent to which this information will impact on one's previous identity and the specific psychological meaning and moral valence (i.e., positive, neutral or negative) it will assume could depend on a number of personal, clinical and socio-cultural factors. In particular, the specific characteristics of an individual's personality structure seem to play a key role: those who lack a secure, stable and integrated sense of self could indeed view a positive test result in terms of an identity threat and feel overwhelmed by the perception of vulnerability arising from the discovery of being a mutation carrier, thus often ending up considering this condition as a confirmation of their sense of inadequacy and defectiveness.

Identity issues could, in turn, have a significant influence on the quality of individual coping mechanisms and shape decisions about treatment and risk disclosure to the family and the wider social context. Specifically, difficulties in incorporating the information of being mutation-positive into one's personal identity could result in the minimization or complete denial of the cancer genetic risk, a tendency often reinforced by a scarce ability to tolerate a state of uncertainty. Clearly, this attitude would lead to underestimating or neglecting the significant clinical implications of a positive test result, as well as a lack of communication of the information to other family members. The inability to handle this knowledge and to adaptively readjust one's self-perception can also induce counselees to put the blame of their mutation on relatives, in particularly parents, who often become vessel of intense feelings of anger and bewilderment.

At the same time, a positive mutation result could offer a sense of relief and validation to individuals who are desperately looking for an explanation which justifies their disease. For example, those who develop cancer despite their healthy habits often consider the diagnosis as a sort of ironic justice and could find it difficult to integrate the disease in their self-perception as "healthy people." In this sense, the discovery of a genetic mutation could help them preserve a sense of identity consistency, by favouring the integration of the illness in a coherent life narrative (Klitzman 2012). They may no longer see the disease as their fault, but as an unescapable genetic trait.

A mutation-positive test result may be comforting also for individuals who tend to consider a cancer diagnosis as a stigma or personal fault, attributing the development of the disease to lack of prevention or negligence of physical symptoms. In these situations, the discovery of a positive test result, by lowering blameworthiness, could help accept the disease and incorporate it into one's identity, thus facilitating adherence to medical treatment and further disclosure to others.

On a more objective level, empirical studies which have investigated the psychological impact of positive genetic test results show that, over time, carriers seem to reach a good level of psychological adjustment. Reviews and meta-analyses indicate that, in general, clinically relevant symptoms of depression, anxiety and distress, as assessed by standardized questionnaires, are rarely detected among mutation-positive counselees (Eijzenga et al. 2014; Hamilton et al. 2009; Hirschberg et al. 2015; Meiser et al. 2005).

However, a more thorough examination of research analysing changes in psychological distress in the post-test phase, both in cancer patients and in non-affected BRCA1 or 2 mutation carriers, reveals the presence of conflicting results. Specifically, several short-term studies have reported an increase in anxiety levels from the first weeks to one month post-testing, with a return to baseline one year after genetic counseling (Bennett et al. 2012; Hamilton et al. 2009; Meiser et al. 2002; van Roosmalen et al. 2004; Watson et al. 2004). Other authors, on the contrary, have not identified any alteration in the counselees' short-term psychological adjustment (Bosch et al. 2012; Claes et al. 2005; Lodder et al., 2001; Reichelt et al. 2004; Ringwald et al. 2016; Schwarz et al. 2002). Long-term studies seem to reveal mixed findings, too: while initial research has reported no differences in psychological outcomes between carriers and non-carriers 3 years and 5 years after testing (Foster et al. 2007; van Oostrom et al. 2003), many authors have later underlined the persistence of elevated levels of distress at 5-years and 7-years follow-up among BRCA1/2 mutation-positive women (Graves et al. 2012; Halbert et al. 2011). These conflicting results could be explained in light of the heterogeneous socio-demographic and clinical characteristics of the samples and the variety of factors involved in psychological adjustment to test result.

Studies which have attempted to identify risk factors for increased distress (Table 1) in the post-test phase have shown that younger age, a lower socio-economic status and a recent cancer diagnosis predict a higher emotional vulnerability to the notification of the positive test result (Esplen and Bleiker 2015). Cancer survivors seem to be more vulnerable to develop emotional distress and feelings of anger than unaffected individuals, as the prospective of a potential relapse might lead them to recall and relive the painful experience of disease and treatment (Di Mattei et al. 2015, 2018). Research findings indicate that also women with children and the first family member to undergo genetic testing are at risk for poor psychological adjustment (Arver et al. 2004; Schlich-Bakker et al. 2008).

Among the psychosocial factors, the pre-test emotional state seems to represent the main predictor of post-test distress (Eijzenga et al. 2014; Hamilton et al. 2009; Meiser et al. 2009). Studies highlight that also a psychiatric history of depression, previous use of psychopharmacologic medication and a higher level of distress at the time of testing significantly increase the risk for distress and anxiety after the communication of the test result (Bosch et al. 2012; Esplen and Bleiker 2015; Lodder et al. 2001; Murakami et al. 2004; Reichelt et al. 2004; van Oostrom et al. 2007). Having experienced a parental loss to hereditary cancer, especially during childhood and early adolescence, and a prior history of trauma or complicated/unresolved grief seem to result in a greater emotional vulnerability, in terms of higher levels of distress, cancer worry and cancer risk perception (Eijzenga et al. 2014; Meiser et al. 2005; van Oostrom et al. 2006; Wellisch et al. 2001).

Additionally, coping style has been identified as another important predictor of post-test distress. Specifically, the use of passive coping strategies, by hampering the assumption of an active role in facing the stressors and in decisions about testing, would result in higher levels of emotional distress after testing (den Heijer et al. 2013; Pieterse et al. 2007; Shiloh et al. 2008). Moreover, avoidant coping seems to be associated with greater psychological vulnerability in the post-test phase compared to problem-oriented ones; indeed, denial and mental detachment of the problem have been identified as risk factors for depressive, anxious and somatization symptoms (Di Mattei et al. 2017, 2018). On the contrary, hopefulness and optimistic attitude have been reported as predictors of future resilience and lower levels of anxiety and depression (Ho et al. 2010). Risk perception represents an additional factor significantly involved in psychological adjustment: several studies underline that the overestimation of cancer genetic risk tend to result in greater distress and cancer worry in the post-test (Cicero et al. 2017; Oosterwijk et al. 2012; Vos, Gómes-García et al. 2012; Vos, Oosterwijk et al. 2012; Voorwinden and Jaspers 2012). In addition, a recent study by Brédart and colleagues (2017), conducted on a sample of breast cancer women, shows that subjective risk perception could moderate the impact of genetic knowledge on emotional distress after the notification of a BRCA1/2 test result. Specifically, authors found that for women who tended to overestimate their cancer risk, the greater genetic knowledge acquired after the pre-test genetic consultation led to increased level of post-test distress. Authors hypothesized that this result might be explained in light of the potential amplification of anxious thoughts and cancer worries prompted by increased knowledge. This evidence has significant clinical implications, suggesting the need to accurately check individual risk perception in evaluating the amount of information to provide and the specific objectives of the pre-test counselling.

Studies indicate that special attention should also be paid to individuals with low levels of social support and high-conflict families, since they seem to be at particular risk for psychosocial problems (van Oostrom et al. 2008).

As discussed earlier, the identification of a pathogenic mutation is usually accompanied by the subsequent implementation of targeted screening programs or risk-reducing interventions, represented by prophylactic mastectomy and salpingo-oophorectomy.

**Table 1. Main risk factors for psychological maladjustment to the notification of a positive test result**

| Socio-demographic factors | Younger age |
| --- | --- |
| | Low socio-economic status |
| | Presence of children |
| Clinical factors | Past cancer experience |
| | Recent cancer diagnosis |
| Psychosocial factors | High level of pre-test distress |
| | Emotional vulnerability at the time of testing |
| | Psychiatric history of depression |
| | Previous use of psychotropic medications |
| | Low self-esteem; identity vulnerability |
| | Prior experience of parental loss to cancer |
| | Prior traumatic experiences; complicated or unresolved grief |
| | Use of passive and avoidant coping style |
| | Overestimation of cancer risk |
| | Low level of social support |
| | Presence of familial conflicts |

Studies which have investigated the psychological impact of surveillance programs for breast cancer have found that mutation-positive women with a family history of cancer do not usually report a significant increase in distress levels nor quality of life impairments (Gopie et al. 2012; Watson et al. 2005).

Ovarian cancer screening (which involves transvaginal ultrasound and CA-125 tests every 4-6 months starting from 30 years) no longer represents a standardized recommendation, since empirical evidence has shown that gynaecological surveillance has not resulted in a significant decrease in ovarian cancer mortality rates (Jacobs et al. 2016). Psychosocial research on this theme indicates that ovarian cancer screening would not determine detrimental consequences on psychological well-being (Brain et al. 2012). However, women who decide to undergo surveillance programs seem to report more cancer worries and a higher perceived risk than those who opt for prophylactic oophorectomy (Madalinska et al. 2005).

With regard to risk-reducing surgery, one must consider that this preventive option could lead BRCA1 or 2 mutation-positive women to confront stressful dilemmas about whether to opt for surgery, when to undergo the operation and how to decide. In general, the choice of prophylactic surgery induces counselees to balance their need to delay, in order to have enough time to come to terms with the psychological implications of the procedure, against the desires to silence cancer worries and prevent the potential development or progression of the disease as soon as possible. Moreover, women have to face uncertainties about the side effects that may result from surgery, concerning not only postsurgical complications, but also potential threats to body image and self-perception, sense of femininity and attractiveness, couple relationships and sexuality. Lack of perceived support from healthcare providers, family and significant others could add complexities to the decision-making process (Klitzman 2012).

Since BRRM and CRRM prophylactic treatments have not shown a clear advantage over intensive surveillance in terms of overall and breast cancer survival, women considering prophylactic mastectomy may tackle particular difficulties in weighing up the burden related to the possible development or progression of the disease against the potential surgical risks. Recent data indicate that relatively low rates of BRCA1 or 2 carriers choose prophylactic mastectomies (0-37%), underlining that the majority of counselees seem to prefer regular surveillance (Botkin et al. 2003; Metcalfe et al. 2005; Schwartz et al. 2012). Systematic reviews suggest that, among psychological factors, the main predictors of the choice to undergo these procedures include fear or worry about the disease, higher perceived vulnerability to future cancer and higher distress and anxiety levels, as well as beliefs that cancer was determined by hormonal or genetic factors (Ager et al. 2016; Braude et al. 2017).

Long-term research investigating the psychological impact of prophylactic mastectomies on both affected patients and high-risk women reports a global positive impact on emotional distress and cancer concern, probably related to the reduction of cancer perceived risk (Geiger et al. 2007; van Oostrom et al. 2003). Furthermore, follow-up studies have highlighted that women tend to show a high overall aesthetic satisfaction six months and one year after prophylactic mastectomy and report no relevant changes in levels of anxiety/depressive symptoms and health-related quality of life (Bai et al. 2019; Brandberg et al. 2012; Gopie et al. 2013, Sahin et al. 2013; Wastenson et al. 2011).

Although the majority of women who undergo BRRM and CPM declare that they are satisfied with their decision (Altschulter et al. 2008; Frost et al. 2011; Haroun et al. 2011; Soran et al. 2015), several studies document how these procedures might produce, especially in affected women, significant impairments in body image perception (Gopie et al. 2013; Heiniger et al. 2015; Metcalfe et al. 2004), sense of femininity, intimate relationships (Esplen and Beiker 2015) and sexuality (Bai et al. 2019; Gaham et al. 2010), which tend to persist for years after surgery. It has been found that concerns about sexual functioning and body image, along with surgical complications, predict an overall lower satisfaction with prophylactic mastectomy, thus indicating a need to pay particular attention to these issues during counselling sessions (Ager et al. 2016; Braude et al. 2017).

RRSO, which actually represents the standard practice in ovarian cancer prevention (Daly et al. 2017) is recommended at the age of 35-40 years in BRCA1 mutation carriers and at 40-45 years in BRCA2 mutation carriers, provided the completion of childbearing desires (NCCN 2016; NICE 2013). The uptake of RRSO has been estimated to be around 49% among high-risk women (Metcalfe et al. 2008). Studies which have investigated psychological factors involved in the decisional process found that high levels of ovarian cancer worry and perceived risk, along with an elevated sense of personal vulnerability to the disease, represent significant predictors of RRSO choice (Gavaruzzi et al. 2017; Halowell et al. 2012; Howard et al. 2011). Moreover, women who worry about transmitting the mutation to offspring and those who feel a sense of family obligation to reduce their cancer risk seem to be more motivated to undergo surgery (Maeland et al. 2014). Likewise, cognitive factors - including awareness about oophorectomy's survival advantages, perceived comprehension of all testing consequences and previous knowledge about BRCA testing - have been documented to predict risk-reducing surgery uptake (Gavaruzzi et al. 2017). Despite its oncological benefits, prophylactic RRSO, which induces a loss of fertility and surgical menopause, may produce relevant physical and psychological sequelae, which could significantly affect health-related quality of life (Altam et al. 2018). Research underlines that physical effects of surgical menopause commonly experienced by oophorectomized women include a wide range of oestrogen deprivation symptoms, such as osteoporosis, stiffness, palpitations, hot flashes, vaginal dryness, sexual pain and discomfort, and poorer physical functioning (Finch and Narod 2011; Michelson et al. 2009; Tucker and Cohen 2017). Women who are premenopausal at the time of surgery tend to report more endocrine symptoms, a greater decline in sexual functioning and a higher level of sexual distress than those who are naturally menopausal (Finch et al. 2012). Although hormone-replacement therapy seems to mitigate endocrine symptoms and improve sexual functioning, many impairments tend to persist despite treatment (Finch et al. 2011; Madalinska et al. 2006). However, studies which have investigated psychosocial issues, although still limited, highlight, in general, a high level of satisfaction with the choice of RRSO (Moldovan et al. 2015) and a favourable impact of this procedure on mental distress, in terms of reduced cancer specific worry, subjective risk perception, depression and anxiety levels (Finch et al. 2013; Madalinska et al. 2006; Michelson et al. 2009).

Given the stressful challenges implied by the decision to undergo risk-reducing surgery and its potential harmful consequences on physical and psychosocial well-being, comprehensive pre-surgical counselling should be mandatory (Braude et al. 2018). This issue is of particular relevance in the context of cancer genetic testing, since the highly subjective nature of decisions about prophylactic surgery often leads providers to feel uncomfortable giving explicit and definitive recommendations, thus leaving women alone in facing this burdensome choice. Contrastingly, some physicians tend to handle these sensitive situations by assuming a directive approach, which could result in providing too premature or constrictive suggestions. These considerations highlight the need to assist women in the decisional-making process, helping them gain a better understanding of the procedure, along with its risks and benefits, share potential doubts and worries and consolidate their personal preferences.

Research seems to confirm the importance of pre-surgical supportive interventions, showing that women who take an active role in the decisional process, driven by adequate information, report significantly higher levels of satisfaction with their choice and a better psychological adjustment to post-surgical challenges (Ager et al. 2016; Braude et al. 2017; Meadows et al., 2018).

## Negative Test Result

The notification of a negative test result is usually accompanied by a sense of relief and a significant reduction in both general anxiety and cancer specific distress. However, for some counselees this good news could come unexpectedly, thus producing feelings of confusion and troubles in psychological adjustment (Bakos et al. 2008; Esplen and Bleiker 2015). This is particularly true for individuals with a strong family history of cancer, who comprehensively develop a high perception of risk and often deeply identify with their relatives affected by the disease. Therefore, a mutation-negative test result compels them to re-adjust their subjective perception of risk to the new knowledge and to face a sense of isolation and alienation from family history. Some individuals seem to find it particularly difficult to deal with this process, which requires the presence of a good level of tolerance to the unexpected, along with the ability to reconsider the personal sense of identity, by integrating the new risk status into one's self-image. Moreover, a negative genetic test could lead individuals who have had previous family experiences of cancer to feel undeservedly saved from a sort of familial destiny, thus bringing out intense feelings of guilt towards the relatives who are affected by the disease and/or carry a pathogenic mutation. The communication of a negative result could reinforce the sense of blame of cancer patients who tend to feel responsible for their disease or think they have contributed in some way to its development (Klitzman 2012). All these aforementioned aspects must be kept in consideration when communicating test results and the individual must be supported during the genetic counseling and testing process.

Empirical research investigating the psychological impact of mutation-negative test results is still limited. A small qualitative study by Macrae and colleagues (2013) analysed psychosocial adjustment in a sample of young women from HBOC families of Ashkenazi Jewish descent who received a true-negative test result. Authors confirmed that the absence of a pathogenic mutation does not necessarily spare counselees experiencing emotional distress, underlining that women reported conflicting emotions of relief, happiness, guilt, fear and anger. Additionally, some of them showed residual cancer worry and, in light of their familial cancer history, expressed a desire to undergo additional screening. This finding seems to be in line with the results of a previous longitudinal study (Claes et al. 2005) which documented the persistence of cancer specific distress both in mutation-carriers and in non-carriers at one-year follow-up.

## Uninformative or Inconclusive Test Results

Uninformative or inconclusive results, that is – negative results in the absence of a mutation previously detected in an affected blood relative, despite the presence of a strong familial breast/ovarian cancer history – imply the possible presence of deleterious mutations in as yet unidentified genes. Research has shown that in these situations the projected lifetime cancer risk, as assessed on the basis of the family pedigree, is significantly higher compared to that of the general population (Metcalfe et al. 2009). As a consequence, the communication of a negative uninformative test result compels counselees to face a situation of uncertainty and to come to terms with highly ambivalent feelings about their genetic risk status, thus representing a potential source of emotional distress. Schroeder and colleagues (2017) conducted a qualitative study, guided by a hermeneutic phenomenological approach, with the objective of gaining a deeper understanding of personal experiences of women living in BRCA-negative families with a familial history of breast cancer. Authors underlined that the knowledge about cancer risk was accompanied by conflicting feelings of restlessness and reassurance. In detail, women described their experiences in terms of a dynamic shifting in and out of normal life spaces, where they get on with their everyday life, trying to maintain the risk awareness in the background, dark "what-if spaces," where fears and anxieties about the potential prospective of the disease come to the foreground, and relational spaces where supportive bonds help them cope with uncertainties.

Despite the relevant psychological implications of inconclusive test results, quantitative research on this theme is still scarce. Available data suggest that the level of distress and cancer worry reported by women who receive an uninformative result seem to be comparable with that of mutation-positive counselees and significantly higher than that of women who tested negative (van Dijk et al. 2008). Studies suggest that personal beliefs and cognitive representations regarding an inconclusive result play a key role in shaping psychological adjustment. Indeed, individuals who report a higher perceived risk and a scarce tolerance of uncertainty tend to develop the highest levels of worry and distress after the notification of the test result (Brédart et al. 2013; O'Neill et al. 2006). These findings have important clinical implications, suggesting the importance to address subjective beliefs and perceptions during counselling sessions to promote psychological well-being.

## Uncertain Test Results

The detection of variants of uncertain clinical significance (VUS) has been consistently increasing since the introduction of whole-genome sequencing and could be particularly challenging both for clients and for clinicians, since these conditions represent a grey, undefined area between positive and negative results.

Communicating a VUS result has been described as putting clients in a sort of "genetic purgatory" (Ackerman 2015), which could raise intense feelings of uncertainty and confusion and lead them to misunderstandings about the meaning and the clinical implications of their situation, thus often determining decisional conflicts about risk management (Hoffman-Andrews, 2018).

Although studies on the psychological impact of VUSs are quite limited, available data seem to suggest that the heightened ambiguity implied by these conditions could result in a significant increase in the level of emotional distress reported by counselees (Maheu and Thorne 2008; O'Neill et al. 2009). Personal beliefs could have a significant influence on psychological adjustment. Specifically, the perception of a higher vulnerability to cancer than objective estimates seems to predict lower levels of anxiety in the post-test, a result probably due to the intense sense of relief generated by the absence of a definitive deleterious mutation (Brédart et al. 2013).

Furthermore, research highlights that individuals who receive a VUS result tend to show greater difficulties in recalling and correctly interpreting this outcome compared to those who obtain a positive, negative or uninformative test result. In particular, Vos and colleagues (2008) found that 67% of individuals who discovered the presence of a VUS in BRCA1 or 2 genes recalled this result as an uninformative outcome and 79% interpreted it as indicating an increase in their genetic risk. Misunderstanding of VUSs seem to be more common in low-educated individuals, thus suggesting the importance to adapt the communication of this information to the specific client characteristics and needs (Richter et al. 2013).

VUS results also force clinicians to face troublesome questions about their disclosure and medical management. Although in clinical practice there is a general consensus on the need to accurately communicate and explain these results, some providers seem to consider not disclosing VUSs under certain circumstances. Moreover, health professionals often show conflicting opinions about the amount of details to provide in communicating this information (Reiff et al., 2013).

Regarding clinical management, the guidelines proposed by the American College of Medical Genetics (ACMG) clearly indicate that variants of uncertain clinical significance should not bring any changes to clients' treatment and follow-up (Richards et al. 2015). Nonetheless, often healthcare providers – in particular those without specific genetic expertise – seem to fail to understand the specific clinical meaning of VUSs and act inappropriately in response to them. A recent study on early stage breast cancer patients with VUSs found that a significant number of breast surgeons reported managing them in the same way as those who are mutation-positive (Kurian et al. 2017).

Another important issue concerns the reclassification of VUSs and the subsequent notification of the new information. Indeed, it is important to consider that the specific nature of these variants (i.e., benign polymorphisms *vs* pathogenic mutations) is usually determined only many years after the initial counselling session, as they require a long period of further intensive research (Murray et al. 2011). This could result in significant difficulties in re-contacting counselees, a task which has been acknowledged by the ACMG as a clinicians' duty (Hirschhorn et al. 1999). Comprehensively, the overwork generated by the management of these situations frequently leads to lack of communication, thus preventing patients acquiring important information about their health, especially in cases in which VUSs are reclassified as pathogenic mutations.

# RISK COMMUNICATION IN CANCER GENETIC COUNSELING

Cancer genetic counseling represents a special process of interaction in the medical field, characterized by a number of specificities that account for its complexity. It could be thought as a "hybrid activity" (Sarangi 2000), which combines the traditional approach of health disciplines with distinctive aspects of counselling professions. Differently from usual medical consultations, focused on providing a diagnosis and an effective treatment, its primary objective is to offer information about risk and support individuals in a difficult decision-making process about the course of action to manage the cancer threat (Chopra and Kelly 2017).

On the objective level, it is to consider that the inherent nature of the information provided in the context of cancer genetic counseling makes its comprehension particularly challenging. Individuals, in fact, are exposed to a large amount of data regarding the purpose of the test, its possible results, the uncertainty of genetic testing, the risk that it reveals unexpected findings and the variety of personal and familial implications of a positive test result. This information is usually conveyed in a complex language, characterized by an abstract and highly specialized terminology, as it may be difficult for providers to translate genetic knowledge into lay terms.

In particular, the communication of cancer risk seems to represent one of the hardest tasks for geneticists during counselling sessions. In oncology, in fact, the notion of risk is far more complex than in classical Mendelian diseases, since cancer risk information involves the consideration of multiple factors (e.g., all the epidemiological, personal and familial risk factors, the probability of having a cancer susceptibility gene running in the family, the probability of developing a sporadic cancer, risks related to cancer prognosis and to preventive/early detection interventions etc.), along with the uncertainty associated with most risk estimators, concerning, in particular, the penetrance values of BRCA genes (Julian-Reyner et al. 2003). The possible detection of uninformative or uncertain test results could add complexities, forcing counselees to come to terms with an unexpected situation and face further ambiguities.

Moreover, subjective factors could interweave with the described aspects, making communication even more critical. Indeed, the feelings of confusion often prompted by genetic testing, the emotional burden related to one's previous familial cancer history and personal life experiences, along with the fears and sense of guilt elicited by the potential transmission of the mutation to offspring could significantly hinder the correct comprehension and subsequent recollection of the information provided. Individual personality characteristics, cognitive biases, personal preconceptions and beliefs may also deeply influence risk interpretation and decisions about cancer prevention and treatment (Dolbeault et al. 2006; Jacobs, Patch and Michie, 2019).

Comprehensively, these issues could lead providers to feel overwhelmed and unable to adequately modulate the interaction to both the informative and the emotional counselees' needs, thus determining difficulties in balancing the necessary professional detachment with emotional closeness and empathic responses.

Literature concerning communication patterns in cancer genetic counseling has identified two predominant models in risk discussion, *id est*, the probability-based approach and the contextual approach (Table 2). The former emphasizes the importance to provide rational data, derived from scientific evidence, which should enable individuals to take an informed decision. According to this approach, risk information should be presented in terms of statistic probability, which can be expressed in several forms, ranging from numbers (absolute or relative risks, odds ratios or ratios), to verbal descriptions of risk magnitude (e.g., "a risk higher than average," "an unlikely event," etc.) and visual displays or graphics (Lipkus and Hollands 1999; Lipkus et al. 1999; Schapira, Nattinger and McHorney 2001). Although the high accuracy of numerical estimates may be appealing for both providers and counselees, research has shown that their impact on individual perception and behavioural intentions cannot be so precisely predicted, as several cognitive and emotional personal issues may alter the interpretation of risk (Lobb et al. 2003; Weinstein 1999). The provision of individual and short-term estimates (rather than population and cumulative risks), along with the conjunction of different information formats, may result in a more accurate perception of genetic risk.

In general, the probability-based approach does not seem to adequately respond to the expectations of counselees, who tend to consider data about risk magnitude as insufficient and report a preference for brief, straightforward and personalised information without statistics (Meiser 2012; Rodin et al. 2009; Rothman and Kiviniemi 1999). Specifically, most individuals express the need to receive clear information about counselling and testing procedure, their own and their relatives' risks and the potential consequences of a positive test result. Other relevant reported needs concern the attainment of detailed information about risk management strategies and emotional support (Pieterse et al. 2005).

In light of these issues, the contextual model highlights the importance to tailor risk communication to the specificities of counselees' individual and familial medical history and assist them in understanding the personal consequences of a given health risk. In line with this approach, it may be effective to share with individuals a hypothesis about potential antecedents of the disease and relevant risk factors involved in its development; another way to contextualise the risk communication process is to increase the personal salience of the risk and illustrate the available preventive interventions, as well as the specific advantages and disadvantages of screening behaviour (Julian-Reyner et al. 2003).

It is important to underline that the two aforementioned models should be considered complementary rather than mutually exclusive, as the integrated use of these approaches may enhance risk communication effectiveness in cancer genetic consultations, helping clinicians convey all relevant rational and standardized information, without losing sight of individual preferences, needs and concerns.

The evaluation of the impact and the effectiveness of risk communication approaches in cancer genetic counseling seems to be particularly challenging, as research in this field presents peculiar limitations which prevent the achievement of consistent results. These are primarily related to the inherent difficulties characterizing experimental research in clinical setting, to the presence of a great heterogeneity of study designs and theoretical backgrounds, as well as to the variety of assessment methods and communication strategies adopted by clinicians. A recent literature review shows that, globally, counselees complain about a lack of attention to psychosocial issues and individual needs (Jacobs et al. 2019). In line with this evidence, several studies have revealed the primarily biomedical and educational nature of cancer genetic counseling, thus underlining that clinical practice in this area does not completely fulfil its goals as a "psycho-educational" process (Biesecker and Peters 2001), aimed at providing information along with psychological support (Resta et al. 2006).

Despite the lack of unanimous consensus about the effectiveness of specific communication strategies, there seems to be a general agreement about the need to implement an individualised multistep approach in cancer genetic counseling, which could offer physicians a guide to effectively shape the interaction with clients through the identification of some core communication principles. Specifically, this approach firstly suggests an accurate assessment of counselees' level of distress, beliefs, preconceptions, preferences, expectations and personal coping style, in order to tailor risk disclosure to individual characteristics and needs. Secondly, clinicians should select and prioritise the information to provide, deciding on the specific content of the communication and the way to convey it. Lastly, counselees should be given a personalized feedback which summarizes the main issues considered during consultations (Julian-Reyner et al. 2003). In particular, written documents which recapitulate information about cancer genetic risk, along with the specific meaning of the test result and the suggested surveillance program and/or risk-reducing measures, have proven particularly effective in alleviating counselees' cognitive burden, thus helping them recall the principal topics discussed with clinicians and take informed decision about their future health (Dolbeault et al. 2006; Jacobs et al. 2019).

**Table 2. Risk-communication models in cancer genetic counseling**

| Probability-based approach | • Standardized communication process<br>• Presentation of objective risk information: use of numerical values (absolute or relative risks, odds ratios or ratios), verbal labels or visual displays and graphics |
|---|---|
| Contextual approach | • Person-centred communication process<br>• Presentation of contextual information to help counselees understand individual risk: concrete exemplification of the potential consequences of risk; illustration of available preventive interventions; discussion of specific advantages and disadvantages of each screening option. |

# SHARING GENETIC INFORMATION WITH FAMILY

The inherent hereditary nature of genetic diseases forces individuals who undergo a mutational analysis to come to terms with the arduous task of disclosing risk information to family members. In the context of cancer genetic counseling, family issues often emerge even before the effective receipt of the test result, as the initial recollection of family history and the process of pedigree construction may induce counselees to discuss with relatives health issues and share previous cancer experiences.

Probands can be viewed as gatekeepers of genetic information, whose intention and ability to communicate about familial risk is crucial in determining the accessibility of this critical information to other family members and their consequent possibility to undertake preventive actions. This "messenger task" (DudokdeWit et al. 1997) may bring out challenging dilemmas of *whether* to tell others and, if so, *whom*, *when* and *how* to communicate about cancer genetic risk and the associated health implications.

Family communication represents a complex dynamic process which leads individuals to weigh the advantages of cancer risk disclosure with the potential harm produced by this information, on both single relatives and extended intra-familial dynamics. Moreover, counselees have to balance their sense of moral responsibility towards kin against conflicting feelings of guilt, shame and fears of possible stigma and misunderstanding (Klitzman 2012).

A great variety of personal, familial, socio-cultural factors and norms could intervene in determining the result of this process and in shaping the specific familial communication patterns. In this sense, a family system perspective may offer a useful theoretical framework in investigating communication issues, helping understand the multiple variables involved in decisions about intra-familial sharing of cancer risk information and appreciate the specificities of interactional dynamics. This approach could, in turn, allow health professionals to integrate the complexities of family systems processes into their clinical practice and recommendations (Galvin and Young 2010).

Specifically, Peterson (2005) has proposed to describe family functioning on the basis of three main dimensions: 1. organization and structure of family relationships; 2. communication processes; 3. collective health-related cognitions and beliefs.

Organization and structure refer to the presence of clear boundaries and family subsystems that define the reciprocal roles and the degree of relational closeness/distance (for example, the subsystem of nuclear family *versus* extended kin may be defined by closer *versus* more distant relations). At the same time, one must consider that the degree of cohesion and reciprocal support could be determined not only by the objective family structure and degree of kinship, but also by shared experiences and personal feelings of intimacy. With respect to this dimension, research investigating cancer risk communication indicates that intra-familial disclosure, as well as genetic testing uptake, tend to be more frequent between relatives who share a close degree of kinship or have intimate bonds (Alegre et al. 2019; Claes et al. 2003; McGivern et al. 2004; Stoffel et al. 2008; Wilson et al. 2004). In particular, studies investigating BRCA1 or 2-cancer related risk highlight that women are more prone to inform female-first degree relatives (Bowen et al. 2004; McGivern et al. 2004; Patenaude et al. 2006; Sermijn et al. 2004), as well as their partners (Forrest et al. 2003; Kenen, Arden-Jones, and Eeles 2004; MacDonald et al. 2007).

Family communication processes denote the specific quality and characteristics of interactional dynamics, which could be deeply shaped by past family history. While mutual trust, reciprocal support and acceptance of a broad range of emotions seem to enhance open dialogues, the presence of family conflicts, emotional barriers and feelings of mistrust may hamper reciprocal exchanges, resulting in an obstructive and ambiguous communication and in the frequent concealment of painful emotions. In line with this, several studies investigating cancer risk disclosure have found that long-standing family tensions, old resentments and hostilities, lack of supportive bonds and perceived emotional distance represent important barriers to intra-familial communication (McGivern et al. 2004; Wilson et al. 2004).

Family health-related cognitions and beliefs, moulded by collective experiences, may also significantly influence disclosure and interpretation of health risk within families. Specifically, empirical evidence has shown that high levels of risk perception, along with increased cancer worries, are significantly associated with the diffusion of genetic information (Chopra and Kelly 2017; Wilson et al. 2004). Further, individuals who have prior knowledge of cancer, related to past personal or familial experiences, seem to be more likely to share cancer risk information with their relatives (Julian-Reyner et al. 2000; Nycum, Avard, and Knoppers 2009). Personal interpretations of genetic test results, as well as their inherent degree of uncertainty, could also shape individual decisions about disclosure. Overall, counselees who tested positive or "true negative" seem to divulge risk information more frequently than those who received uninformative or uncertain results, whose meaning could be confusing and hard to explain (Chopra and Kelly 2017; McGivern et al. 2004; Wagner Costalas et al. 2003).

While the understanding of the multiple issues involved in the decision to disclose risk genetic information has become increasingly extensive, research analysing the specific content of this communication and how and when it occurs is still quite limited.

Nonetheless, it is important to note that counselees often face troublesome questions, not only about whether to disseminate risk information, but also about the exact way to handle this communication. Primarily, it is necessary to define the specific amount of information to convey and its level of specificity, which may range from partial and vague or superficial divulgations to complete and highly detailed discussions. Further, "messengers" have to take complex decisions about how to tell their families - evaluating whether to reveal information directly and explicitly or in a more gradual, implicit and indirect way - as well as about the most appropriate time frame for sharing genetic information (Klitzman 2012). The result of this multi-faceted decisional process, which could be influenced by a variety of individual and relational factors, closely depends on the specific recipients of the information and their anticipated emotional reactions to disclosure.

In particular, communication to offspring appears to be one of the most challenging aspects of genetic testing process and represents a key transformative and translational moment in the family system life, bringing awareness to another generation of hereditary risk.

In discussing test result with children, parents have to face highly conflicting emotions and balance their desire to protect offspring from the emotional impact of risk disclosure against the awareness about the importance of this information for their future health.

The communication of cancer hereditary risk to offspring ultimately involves questions about the well-being of the parents and the future generations, as well as the continuity of the parent-child relationship. The threat to attachment bonds evoked by the diffusion of genetic information may consequently bring out deep emotions of fear, anxiety, sorrow and anger, which could significantly hamper open and clear dialogues. Specifically, an in-depth exploration of parental concerns has revealed that fears are often strictly interwoven with feelings of guilt about transmitting mutations and encumbering children with overwhelming information. Parents' fears may also concern the ability to talk to children about the test result and its implications without transmitting excessive grief and worries; this issue could be particularly relevant in discussing cancer genetic risk with younger children, who tend to rely predominantly on parental affect in interpreting the seriousness of the communication. Moreover, a remarkable number of parents seem to perceive their children as extremely vulnerable and emotionally frail, thus worrying that the discussion of the test result will evoke intense feelings of anguish, which they might not be ready to handle (Patenaude and Schneider 2017).

Further, worries may arise around the possibility that the knowledge of the parental and personal risk status might interfere with children's healthy developmental process and their adjustment to growth changes. This issue mainly emerges in communication to adolescents and becomes especially critical in mother-daughter discussions (Fisher et al. 2014). With respect to this topic, research has shown that the hesitation of some mothers to communicate about BRCA1 or 2 mutations is often motivated by the fear that the discovery of breast and ovarian cancer risks could undermine their daughters' acceptance of pubertal body changes (Patenaude and Schneider 2017).

Fears of interfering with important life process, like applying to college and searching for a job, or with meaningful children's experiences, such as graduation, engagement or marriage and child birth, may represent further barriers to disclosure.

Parent-child communication of cancer genetic risk could make decisions about what and when to relay it particularly challenging: evaluating the specific content of the disclosure and identifying the right moment and age to inform children could, in fact, bring out difficult questions about children's levels of maturity and their ability to understand risk information. With respect to these issues, one must consider that parental perception of children's ability to handle genetic risk knowledge and cope with its implications often results in it being distorted by personal needs to minimize or deny the risk and desires to protect one's offspring (Klitzman 2012).

The presence of children of different ages could add complexities, forcing counselees to adapt the communication to specific individual characteristics and needs. Undoubtedly, disclosure of BRCA1 or 2-cancer related risk should be prioritised towards daughters who are close to the age recommended to begin breast screening. In these situations, separate discussions may allow daughters to receive more extensive explanations about the meaning of the test result and its potential effects on health. Moreover, private dialogues could enable them to express their concerns and emotional reactions, which are usually stronger than those experienced from younger children, for whom the information tends to assume a more abstract meaning (Wong et al. 2010). However, the complete exclusion of younger family members from family discussions may burden the older ones with a difficult secret-keeping task, which could, in turn, hinder intra-familial interactions and communication processes.

Difficult quandaries may also emerge about communication of cancer genetic risk to extended family members, from cousins, nieces and nephews to "long-lost" relatives. In fact, in these situations the sense of moral duty and personal responsibility to divulge cancer risk information is usually less intense and urgent compared with nuclear family, since relationships with extended kin tend to be more fluid or superficial and less socially prescribed. Geographical distances, lack of intimate bonds, limited knowledge about distant relatives and concerns about their potential reactions to risk disclosure could further obstruct communication (Klitzman 2012).

The numerous challenges involved in family communication clearly suggest that counselees need a specific professional guidance in navigating conversations about cancer hereditary risk. Ideally, clinicians and genetic counsellors should prepare clients to address communication issues, helping them consider in advance *whether, what, how* and *when* to disclose genetic information and wrestle with difficult dilemmas and highly conflicting emotions. Specifically, it is crucial to accurately explore fears, feelings of guilt, distorted cognitive representations and beliefs which may hamper open communication, thus preventing the diffusion of important information for family members' health and their consequent undertaking of preventive measures. In investigating communication barriers, health professionals should adopt a respectful and non-judgmental approach, in order to allow counselees to feel comfortable enough to share their emotional experience.

Special attention should be paid to the multiple complexities involved in parent-child communication. Health providers should assist parents in considering the benefits of telling, while exploring their fears about the potential harm of genetic information. Further, helping them acknowledge the actual child's developmental level, along with their cognitive and emotional resources, may be crucial in mitigating their sense of guilt about transmitting excessively burdensome information. In this sense, parents may also feel relieved by considering that the initial talk to their children is only the starting point of a long-term conversation, which is expected to become increasingly in-depth over time. In parallel, addressing parents' feelings of inadequacy may be vital to helping them appreciate their own strengths and abilities to handle conversations about cancer hereditary risk, along with children's emotional reactions.

Comprehensively, the exploration of these communication issues represents a delicate and time-consuming task, which may require a specific professional training. Clinicians and genetic counsellors might not have specialized competence in this field or find it difficult to address communication problems, as the increasingly demanding client informational needs seem to leave limited time for other counselling issues. These considerations suggest the importance of including trained psychologists in cancer genetic counseling, who may offer specific psychosocial support and, if necessary, more in-depth psychotherapeutic interventions. The possibility to refer counselees to specialized mental health professionals may help relieve the growing burden of genetic counsellors, thus improving the global quality of cancer genetic services.

## PSYCHOLOGICAL SUPPORT DURING GENETIC TESTING

As we have seen in the previous paragraphs, undergoing genetic testing for cancer, and receiving the results can be a distressing process and can have serious psychological consequences. Mental health care and support should be integrated into the standard genetic testing and counselling process in order to aid psychological adjustment and give individuals and their caregivers support.

The responsibility of assisting and supporting the client on both psycho-emotional and psycho-social levels should be ascribed to a psychologist or genetic counsellor who encourages and facilitates the support from the counselee's various social spheres (family, work friends etc.). The psychologist's role and duty should be activated from the very first phases of counselling in direct and integrated collaboration with geneticists and oncologists, and not as a subsequent and additional requested action. As a result, the psychologist can connect and create a relationship with the client since the preliminary phases of the process, phases that would help to avoid and eliminate the social stigma tied to a wrongful cultural vision and perception of the psychologist's profession. This would also ease and improve the alliance between the psychologist and client, and consequently also facilitate the sharing of the information linked not only to the genetics of the disease but also to the history of both the client and her family. The combined intervention of doctors and psychologists helps to develop a feeling of acceptance and empathy towards the client, so the latter feels regarded by the professionals as a person, and not only as a mere object of study and testing. Besides this, it might enhance the feelings of being treated with respect, perceiving that the professionals are dedicating and devoting their care and time to the individual's case.

To begin with an individual's personal, medical and cultural history must be collected. These are usually obtained by a genetic counsellor or psychologist. For women who could be at risk of serious psychological issues, or who experience challenges with family communication or decision-making, the ongoing mental health support is crucial. Thus, baseline levels of anxiety and distress, any vulnerabilities or possible psychopathologies, both specific and generalized, should be investigated, aiming to prevent possible serious psychological implications that might evolve into psycho-pathologies of different entities later on in the process. Psychologists will also evaluate which are the possible resources upon which it is successively possible to focus on and to develop, in order to promote and support an individual's self-esteem, first and foremost, and a support oriented to sustain the required work that ought to be done in the following sessions. Self-esteem, mastery, personal power, control, sense of guilt, shame and anything else that might erode self-esteem itself should also be explored. Furthermore, emotions, cognitions, beliefs, needs, concerns, expectations and hopes should also be noted in order to be strengthened or re-evaluated later on (Farina and Crimi 2017).

Psychological screening instruments are useful and have been structured, designed and tested for at-risk populations (Esplen et al. 2011; Kasparian et al. 2007). These standardized tools - such as the Multidimensional Impact of Cancer Risk Assessment (Cella et al. 2002), the Psychological Adaptation to Genetic Information Scale (Read, Perry, and Duffy 2005), the Genetic Risk Assessment of Coping Evaluation (Phelps et al. 2010) and the Psychosocial Aspects of Hereditary Cancer (Eijzenga et al. 2014) - help us identify risk markers related to not only the cancer experience but also to the genetic testing process. Risk factors that can be pinpointed during the initial evaluation stage include three main categories: socio-demographic, medical and psychological risk factors. Socio-demographic factors include: age, gender, culture or ethnicity, socioeconomic status, having young children and education. Medical risk factors include: penetrance, severity/nature of the disease, prevention options and risk-reducing procedures. Lastly, psychosocial risk factors include: loss of a relative to the disease, having a family member with the disease, having a history of other life traumas or losses, recounting a morbid psychological history or condition, current psychological functioning (presence of anxiety, depression or disease-specific worry), other current life stressors, expecting to receive a negative result, coping style and social support level (Esplen, 2006).

A preliminary evaluation of all emotional, psychological, psycho-social aspects allows the multidisciplinary team of health professionals to obtain an estimate and evaluation of the possible number of genetic counseling sessions the client might need. Furthermore, it would provide information which is useful in obtaining a preliminary estimate and evaluation about the need of additional sessions, specifically for the support concerning psychological, emotional and/or psychosocial issues. Due to the numerous personal differences pertaining to each client, it is important to evaluate each individual's functioning and representations in order to improve personal well-being. Sometimes, the genetic counseling and testing process is too distressing and could bring about more negative aspects for a client and, in this case, it can be postponed until the client is ready to face the genetic testing process.

Therefore, a psychologist should consider and run through all those processes meant to promote health that are connected to the quality of life and are meant to support the so-called "life skills" of the individual. The task and objective of the psychologist is to instil in both the counselee and their relatives the belief and certainty that they have the capability of positively modifying their habits by putting into effect appropriate and functional behaviours to improve health and well-being; this means that the clients themselves have power of self and collective efficacy over their own well-being (Farina and Crimi 2017).

**Table 3. Recommended psychosocial interventions for specific level
of emotional distress (Adapted from Esplen 2006)**

| LOW DISTRESS LEVEL | MODERATE DISTRESS LEVEL | HIGH DISTRESS LEVEL |
|---|---|---|
| • Psychoeducational interventions and informative material (i.e., leaflets, Internet websites)<br>• Decisional aids<br>• Peer support | • Stress-management interventions and coping strategies (e.g., CBT)<br>• Additional counselling sessions<br>• Individual or group psychosocial support | • Individual psychotherapeutic interventions (e.g., CBT or psychodynamic)<br>• Psychiatric assessment, with possible implementation of psychotropic medication |

Once the individual's background and level of distress have been examined and the psychologist feels there is a need to modify certain aspects of a client's cognition, emotional experience or expectations, it is helpful to administer an intervention, be it psychosocial or psychotherapeutic. Current research mostly focuses on psycho-educational methods and decision-making aids. Esplen and colleagues (2006) stated that interventions can be chosen along a continuum using the level of distress communicated by the client (Table 3). For instance, individuals who experience low distress levels can be given direct and simple educational materials such as pamphlets, CDs, internet information, peer support, in-person or telephone-based counselling approaches (Appleton et al. 2004; Graves et al. 2010; Lerman et al. 1995; Miller et al. 2005; Wellisch et al. 1991) and decisional aids (Miller et al. 2005, Wakefield et al. 2007). These have shown to improve knowledge and coping with the genetic testing process. Individuals with moderate levels of anxiety or individuals who are experiencing sleep disruption, difficulty with decision-making and communication challenges will benefit from additional one-to-one counselling sessions with a genetic counsellor or mental healthcare professional. They may also benefit from individual or group peer support and cognitive-behavioural strategies to manage stress and facilitate tolerance of ongoing uncertainty (Wellisch et al. 1999). When there is a high level of distress (depressive and anxiety symptoms, hopelessness or suicidal ideation) in individuals, more specialized services should be offered by a mental healthcare professional. Individual psychotherapy and psychotropic medication may help to address unresolved grief and alleviate symptoms. Furthermore, professionally led support groups and longer follow-up may also be undertaken (Esplen et al. 2004; Wellisch et al. 1999).

## Cognitive and Behavioural Interventions

Cognitive-behavioural interventions are used often in healthcare settings to lower anxiety levels and facilitate stress management. The main idea behind this approach is that there are deep-rooted links between our thoughts, behaviours and emotions. What one thinks has an influence on how one acts and feels. How one feels, in turn, affects how one thinks and acts, and how one acts reflects how one thinks and feels. The majority of Cognitive Behavioural Therapies (CBT) focuses on cognition as the point of intervention, to effect change in emotions and behaviour (Biesecker et al. 2017). This is especially important when considering application of CBT to genetic counseling, with its implicit connection between medical education and psychotherapeutic counselling. CBT guides the practitioner to focus on what the individual *"thinks"* and *"believes,"* as it is the client's thoughts that shape the emotional experience (Biesecker et al. 2017).

To address an over-inflated cancer risk, re-framing (an intervention frequently used in genetic counseling which has its origin in CBT) can be used. When genetic counsellors use person-first language or re-state the risk of a mutation, the goal is to affect how the individual thinks about their condition. In this way a "dysfunctional" or "catastrophic" thought can be re-structured. CBT techniques that can help to facilitate coping with genetic information include finding and challenging what are referred to in CBT as "distorted thoughts" and "irrational beliefs." These are thoughts and credences that conflict with reality and are experienced by all of us, particularly in response to stressors. Such thoughts are frequently the source of emotional distress or maladaptive behaviour and if we work to eliminate them, this can lead to improved emotional and behavioural outcomes. On the contrary, focusing intentionally on adaptive cognitions can also be beneficial to client outcomes. As this work involves confronting a client's beliefs, it requires a strong therapeutic relationship (Biesecker et al. 2017).

Exercises such as keeping a diary and writing down thoughts and ideas can help the counsellor to gain insight into specific beliefs and rigid thought patterns and they can be communicated to the individual so that they are more aware of their distorted cognitions and can encourage more realistic interpretations of their situation. CBT can also help individuals adopt preventive healthcare strategies and adhere to screening programmes (Grassi and Riba 2012).

An example of a CBT approach that has been used with women who underwent mastectomies for breast cancer demonstrated that the intervention group who received consultation based on Ellis rational emotive behavior therapy (a type of CBT which is a short-term psychotherapy that helps you identify self-defeating thoughts and feelings, challenge the rationality of those feelings, and replace them with healthier, more productive beliefs) for 6 sessions during 3 weeks had a significantly lower body image score (Fadaei et al. 2011). This study emphasizes the importance of offering consultation in breast cancer patients.

Furthermore, relaxation and distraction techniques can help to address stress associated with repeated medical and screening appointments (Phelps et al. 2006). Specific programmes such as mindfulness-based stress reduction training have been analysed and are effective in improving quality of life and reducing anxiety in cancer patients (Carlson et al. 2004). Thus cognitive-behavioural strategies seem to be useful in supporting individuals throughout the genetic counseling process from the pre-test phase to the post-test phase and also in follow-up.

## Psychodynamic Approaches

This approach emphasizes emotions more than cognitions and uses the relationship between the therapist and individual as a way to understand the individual and create change. Insight is gained by looking at early life experiences such as previous losses or traumas (particularly those related to cancer) and seeing how they relate to the current distress. Once self-awareness is established, the emotional patterns can be explored and challenged through the comprehension of how they developed.

Different aspects can be worked on, such as "attachment style." This refers to the basic beliefs and behaviours that were formed in childhood about relationships and which persist in adulthood for most people (Bowlby 1988; Maunder and Hunter 2008). This approach can be helpful in gaining a comprehension of one's motivations and behaviours. For example, attachment style may help to predict a person's ability to trust others, regulate anxiety, process fearful situations and tolerate the unknown; all of these aspects pertain to the genetic counseling situation.

Usually "defense mechanisms," which operate on an unconscious level in order to ward off unpleasant feelings, will be evaluated. For instance, an individual's use of negation and minimization will be verified, and the psychologist will have to judge whether these are adaptive or not in protecting the individual from a low self-esteem and self-control.

Thus, defenses could possibly be approvable and supportive, or, on the contrary, if they are maladaptive, they have to be reduced and, if possible, modified in order to reduce the development of further psychopathologies and increased stress levels during the genetic counseling process (Di Mattei et al. 2015, Farina and Crimi, 2017).

Other aspects that are usually always present in a psychodynamic approach are: creating an atmosphere of mutual understanding and allowing individuals to express their emotions in a safe setting where trust can be formed for the processing of unresolved issues (Okabayashi 2017).

An example of a psychodynamic approach used in a medical setting is described in a study by Beutel and collegues (2014), who wanted to determine the efficacy of a short-term psychodynamic treatment in breast cancer patients who were also diagnosed with depression using the Hospital Anxiety and Depression Scale. They found that the intervention group who underwent a Short-Term Psychodynamic Psychotherapy (STPP) achieved significantly more remission than the control group. They concluded that STPP is an effective treatment of a broad range of depressive conditions in breast cancer patients improving depression and functional QoL.


## Group Approaches

Group counselling occurs when various individuals have genetic counseling sessions together, typically for the same indication (e.g., all have a family history of breast cancer) (Cohen et al. 2012), in the pre-test phase, during the testing and post-test phase; it is typically performed in person.

Groups have shown improvements in decreasing cancer worry and general anxiety, and also improving coping strategies (Esplen et al. 2004; Wellisch et al. 1999). They help to facilitate decision-making around genetic testing or prevention, as well as assisting with family communication (Esplen et al. 2004, Wakefield et al. 2007). For example, the Socio-Psychological Research in Genomics (SPRinG) Collaboration have developed a new intervention, based on multi-family discussion groups (MFDGs), to support families affected by inherited genetic conditions and train genetic counsellors in its delivery. MFDGs are integrative interventions that use a variety of concepts including systemic therapy, CBT and group therapy practices. They involve working with between half a dozen families at the same time. Families who share an experience are assembled together and facilitated by therapists, to explore the issues they need to confront and in order to discover their family's strengths in dealing with those issues. This increases the family's sense of identity and self-esteem, reduces a sense of isolation and stigmatisation and helps them to build and maintain strong supportive relationships (Asen and Scholz 2010). Eisler et al. (2017) found that MFDGs are a potential useful resource in supporting families to communicate genetic risk information and can enhance family function and emotional well-being. Moreover, they demonstrated that it is feasible to train genetic counsellors in the delivery of the intervention and that it has the potential to be integrated into clinical practice. Its longer-term use in routine clinical practice, however, relies on changes in both organisation of clinical genetics services and genetic counsellors' professional development.

SEGT (supportive-expressive group therapy) (Spiegel and Spira 1991) has been applied to the cancer genetic counseling process. This group approach brings together a handful of individuals with common challenges and issues (e.g., carrying a genetic mutation) and via 90 minute sessions which are conducted weekly for at least 8 weeks (and then monthly thereafter) aims to use the existential impact of the genetic situation as an opportunity to explore its relevance in relation to perceptions of cancer risk (Esplen et al. 1998; Esplen et al. 2000; Esplen et al. 2004). It does this by exploring past experiences concerning family and cancer and their influence on current risk processing, opinions and emotional impact, so as to help with decision-making and adjustment. Role modelling of others in the group and the learning that occurs through these shared experiences facilitate coping and legitimize the emotions an individual may feel in this process (Spiegel and Spira 1991). SEGT has been tested in women at risk of developing cancer and with those who carry mutations. It demonstrated improvement in coping, grief and quality of life (Esplen et al. 2000; Esplen et al. 2004; Spiegel and Spira 1991). Moreover, including women who have survived cancer in the group may also provide an opportunity for direct exchange between these two groups of women. Young women may be reassured by the fact that older women who have survived cancer are present in the group, thus challenging their belief that death is inevitable.

Group genetic counseling has been used by up to 10% of cancer genetic counsellors, however, it is rarely the sole service delivery model used by a counsellor (Cohen et al. 2013; Trepanier and Allain 2014). Group genetic counseling has shown promise for increasing efficiency by decreasing per-patient time for genetics clinicians (Ridge et al. 2009). Nonetheless, questions remain about whether group counselling would be widely accepted by cancer genetic counseling clients. One study showed a high rate of declining group counselling, concerns about the effects of group dynamics on client privacy and decision-making, and a preference for individual counselling over group counselling (Ridge et al. 2009). A later, non-randomized study echoed this preference for individual counselling when clients were given the choice of service delivery model (Rothwell et al. 2012). We believe this model can certainly be integrated into an individual therapy plan. Using alternate service delivery models to provide cancer genetic counseling involves balancing several factors thought to be important to the clinical experience, including clients' access to care and clinicians' perceptions of their own effectiveness to clearly explain potentially complex genetic concepts, while at the same time assessing and responding to psychosocial cues. Support models ultimately need to strike this balance while maintaining client confidentiality, fitting into the healthcare systems' work flows, and being financially viable.

## CONCLUSION

In summary, within this chapter we have laid out what hereditary gynaecological cancers are, the genes which are most often involved, the risks of developing cancer and possible prophylactic surgeries or chemopreventative treatments that can reduce the risk of developing breast or ovarian cancer. We then discussed the genetic testing and counselling procedures from the pre-test phase all the way to follow-up, all the while considering the test result, that is, positive, negative, inconclusive or uncertain. We also presented the clinical advantages offered by the treatment-focused genetic testing, which represents a new developed, individualised approach to the testing and counselling process, aimed at enhancing patient-centred cancer care. Next, we turned to more personal aspects of counselling and testing, *id est,* decision-making, especially in the context of having other family members who may also need to undergo testing or at least be aware of their risks. From this, we went on to discuss the psychological impact of genetic testing, keeping in mind the different result options, and we discussed the emotional reactions and psychological adjustments which may vary widely from pre-test all the way through to follow-up. Furthermore, we discussed the long-term impact of the genetic testing procedure on counselees and counsellors. We therefore tried to highlight the dynamic and multifactorial nature of the adjustment process to cancer genetic testing, considering its possible changes over time and the multiple personal, medical, familial and social factors which may interweave in shaping individual emotional responses.

We considered communication issues that may arise during the testing process and the fact that clients may need professionals who have specific training in this field so that they can discuss their fears and problems. Later, we went on to talk about the different psychotherapeutic approaches that are possible when supporting a counselee, for example, cognitive-behavioural, psychodynamic and group therapy; these are useful to help the client during times of crisis, in decision-making and also to help improve their quality of life.

In conclusion, there seems to be a general consensus about the need to implement individualised multi-step approaches during cancer genetic testing and counselling: no one individual is the same as another and so each intervention must be adapted and tailored to the client's needs and to their specific medical situation. We hope that we have shed some light on the most recent research available regarding the psychological effects of cancer genetic testing and counselling.

# REFERENCES

Ackerman, M. J. (2015). Genetic purgatory and the cardiac channelopathies: exposing the variants of uncertain/unknown significance issue. *Heart rhythm,* 12: 2325-2331.

Ager, B., Butow, P., Jansen, J., Philips, K. A., Porter, D. and CPM DA Advisory Group. (2016). Contralateral prophylactic mastectomy (CPM): a systematic review of patient reported factors and psychological predictors influencing choice and satisfaction. *The Breast,* 28: 107-120.

Algegre, N., Vande Perre, P., Jean Bignon, Y., et al. (2019). Psychosocial and clinical factors of probands impacting intrafamilial disclosure and uptake of genetic testing among 103 French families with BRCA 1/2 or MMR gene mutations. *Psycho-Oncology,* 1-8.

Alsop, K., Fereday, S., Meldrum, C., et al. (2012). BRCA mutation frequency and patterns of treatment response in BRCA mutation–positive women with ovarian cancer: a report from the Australian Ovarian Cancer Study Group. *Journal of Clinical Oncology,* 30: 2654.

Altman A. M., Hui, J. Y. C. and Tuttle, T. M. (2018). Quality of life implications of risk-reducing cancer surgery. *British Journal of Surgery,* 105: e121-e130.

Altschuler, A., Nekhlyudov, L., Rolnick, S. J., et al. (2008). Positive, negative, and disparate? Women's differing long-term psychosocial experiences of bilateral or contralateral prophylactic mastectomy. *The Breast Journal,* 14: 25-32.

American Society of Clinical Oncology. (2003). American Society of Clinical Oncology policy statement update: genetic testing for cancer susceptibility. *Journal of Clinical Oncology,* 21: 2397-2406.

Appleton, S., Watson, M. and Rush, R. (2004). A randomised controlled trial of a psychoeducational intervention for women at increased risk of breast cancer. *British Journal of Cancer,* 90: 41-47.

Arindem, P. and Soumen, P. (2014). The breast cancer susceptibility genes (BRCA) in breast and ovarian cancers. *Frontiers in Bioscience,* 19: 605-618.

Arver B., Haegermark, A., Platten, U., Lindblom, A. and Brandberg, Y. (2004). Evaluation of psychosocial effects of pre-symptomatic testing for breast/ovarian and colon cancer pre-disposing genes: a 12-month follow-up. *Familial Cancer,* 3: 109-116.

Asen, E., and Scholz, M. (2010). *Multiple-family therapy: concepts and techniques.* East Sussex: Routledge.

Augestad, M. T., Høberg-Vetti, H., Bjorvatn, C. and Sekse, R. J. T. (2017). Identifying needs: a qualitative study of women's experiences regarding rapid genetic testing for hereditary breast and ovarian cancer in the DNA BONus study. *Journal of Genetic Counseling,* 26: 182-189.

Backes, F. J, Mitchell, E., Hampel, H. and Cohn, D.E. (2011). Endometrial cancer patients and compliance with genetic counseling: room for improvement. *Gynecologic Oncology,* 123: 532-536.

Bai, L., Arver, B., Johansson, H., Sandelin, K. and Wickman, M. (2019). Body image problems in women with and without breast cancer 6-20 years after bilateral risk-reducing surgery – a prospective follow-up study. *The Breast,* 44: 120-127.

Bakos, A. D., Hutson, S. P., Loud, J. T., Peters, J. A., Giusti, R. M. and Greene, M. H. (2008). BRCA mutation-negative women from hereditary breast and ovarian cancer families: a qualitative study of the BRCA-negative experience. *Health Expectations,* 11: 220-231.

Bennett, P., Phelps, C., Hilgart, J., Hood, K., Brai, K. and Murray, A. (2012). Concerns and coping during cancer genetic risk assessment. *Psycho-Oncology,* 21: 611-617.

Biesecker, B., Austin, J. and Calesghi, C. (2017). Theories for Psychotherapeutic Genetic Counseling: Fuzzy Trace Theory and Cognitive Behavior Theory. *Journal of Genetic counseling,* 26: 322-330.

Biesecker, B. B., and Peters, K. F. (2001). Process studies in genetic counseling: peering into the black box. *American Journal of Medical Genetic,s* 106:191-198.

Bleiker, E., Wigbout, G., van Rens, A., Verhoef, S., Van't Veer, L. and Aaronson, N. (2005). Withdrawal from genetic counseling for cancer. 2005. *Hereditary Cancer in Clinical Practice,* 3: 19-27.

Bosch, N., Junyent, N., Gadea N., et al. (2012). What factors may influence psychological well being at three months and one year post BRCA genetic result disclosure? *The Breast,* 21: 755-760.

Botkin, J. R., Smith, K. R., Croyle, R. T., et al. (2003). Genetic testing for a BRCA1 mutation: prophylactic surgery and screening behavior in women 2 years post testing. *American Journal of Medical Genetics Part A,* 118: 201-209.

Bowen D. J., Bourcier, E., Press, N., Lewis, F. M. and Burke, W. (2004). Effects of individual and family functioning on interest in genetic testing. *Journal of Community Genetics,* 7: 25-32.

Bowlby, J. (1988). *A Secure Base: Clinical Applications of Attachment Theory.* London: Routledge.

Brain, K. E., Lifford, K. J., Fraser, L., et al. (2012). Psychological outcomes of familial ovarian cancer screening: no evidence of long-term harm. *Gynecologic Oncology,* 127: 556-563.

Brandberg, Y., Arver, B., Johansson, H., Wickman, M., Sandelin, K. and Liljegren, A. (2012). Less correspondence between expectations before and cosmetic results after risk- educing mastectomy in women who are mutation carriers: a prospective study. *European Journal of Surgical Oncology (EJSO),* 38: 38-43.

Braude, L., Kirsten, L., Gilchrist, J. and Juraskova, I. (2018). The development of a template for psychological assessment of women considering risk-reducing or contralateral prophylactic mastectomy: a national Delphi consensus study. *Psycho-Oncology*, 27: 2349-2356.

Braude, L., Kirsten, L., Gilchrist, J. and Juraskova, I. (2017). A systematic review of women's satisfaction and regret following risk-reducing mastectomy. *Patient Education and Counselling,* 100: 2182-2189.

Brédart, A., Kop, J. L., DePauw, A., et al. (2013). Short-term psychological impact of the BRCA1/2 test result in women with breast cancer according to their perceived probability of genetic predisposition to cancer. *British Journal of Cancer,* 108: 1012-1020.

Brédart, A., Kop, J. L., De Pauw, A., Caron, O., Fajac, A., Nogue, C., Stoppa-Lyonnet, D. and Dolbeault, S. (2017). Effect on perceived control and psychological distress of genetic knowledge in women with breast cancer receiving a BRCA1/2 test result. *The Breast,* 31: 121-127.

Bresser, P. J., Seynaeve, C., Van Gool, A. R., et al. (2007). The course of distress in women at increased risk of breast and ovarian cancer due to an (identified) genetic susceptibility who opt for prophylactic mastectomy and/or salpingo-oophorectomy. *European Journal of Cancer*, 43: 95-103.

Buchanan, A. H., Voils, C. I., Schildkraut, J. M., et al. (2017). Adherence to Recommended Risk Management among Unaffected Women with a *BRCA* Mutation. *Journal of Genetic counseling,* 26: 79-92.

Carlson, L. E., Speca, M., Patel, K. D. and Goodey, E. (2004). Mindfulness-based stress reduction in relation to quality of life, mood, symptoms of stress and levels of cortisol, dehydroepiandrosterone sulfate (DHEAS) and melatonin in breast and prostate cancer outpatients. *Psychoneuroendocrinology,* 29: 448-474.

Caruso A., Vigna, C., Maggi, G., Scga, F. M., Cognctti, F. and Savarese, A. (2008). The withdrawal from oncogenetic counseling and testing for hereditary and familial breast and ovarian cancer. A descriptive study of an Italian sample. *Journal of Experimental & Clinical Cancer Research,* 27: 75.

Chopra, I., and Kelly, K. M. (2017). Cancer risk information sharing: the experience of individuals receiving genetic counseling for BRCA1/2 mutations. *Journal of health communication,* 22: 143-152.

Cicero, G., De Luca, R., Dorangricchia, P., Lo Coco, G., Guarnaccia, C., Fanale, D., Calò, V. and Russo, A. (2017). Risk perception and psychological distress in genetic counseling for hereditary breast and/or ovarian cancer. *Journal of Genetic counseling,* 26: 999-1007.

Claes, E., Evers-Kiebooms, G., Boogaerts, A., Decruyenaere, M., Denayer, L. and Legius, E. (2003). Communication with close and distant relatives in the context of genetic testing for hereditary breast and ovarian cancer in cancer patients. *American Journal of Medical Genetics,* 116A: 11-19.

Claes, E., Evers-Kiebooms, G., Decruyenaere, L., Boogaerts, A., Phillip, K. and Legius, E. (2005). Predictive genetic testing for hereditary breast and ovarian cancer: psychological distress and illness representations 1 year following disclosure. *Journal of Genetic Counseling,* 14: 349-363.

Cohen, S. A., Marvin, M. L., Riley, B. D., Vig, H. S., Rousseau, J. A. and Gustafson, S. L. (2013). Identification of genetic counseling service delivery models in practice: a report from the NSGC Service Delivery Model Task Force. *Journal of Genetic counseling,* 22: 411-421.

Cohen, S. A., Gustafson, S. L., Marvin, M. L., Riley, B. D., Uhlmann, W. R., Liebers, S. B. and Rousseau, J. A. (2012). Report from the National Society of Genetic Counselors service delivery model task force: a proposal to define models, components, and modes of referral. *Journal of Genetic counseling,* 21: 645-651.

Crook, A., Plunkett, L., Forrest, L. E., et al. (2015). Connecting patients, researchers and clinical genetics services: the experiences of participants in the Australian Ovarian Cancer Study (AOCS). *European Journal of Human Genetics,* 23: 152-158.

Culver, J., Burke, W., Yasui, Y., Durfy, S. and Press, N. (2001). Participation in breast cancer genetic counseling: the influence of educational level, ethnic background, and risk perception. *Journal of Genetic Counseling,* 10: 215-231.

Cuzick, J., Sestak, I., Forbes, J. F., et al. (2014). Anastrozole for prevention of breast cancer in high-risk postmenopausal women (IBIS-II): an international, double-blind, randomised placebo-controlled trial. *The Lancet,* 383: 1041-1048.

Daly, M. B., Pilarski, R., Berry, M., et al. (2017). NCCN guidelines insights: genetic/familial high-risk assessment: breast and ovarian, version 2.2017. *Journal of the National Comprehensive Cancer Network,* 15: 9-20.

Dekker, N., Dorst, E., Luijt, R., et al. (2013). Acceptance of genetic counseling and testing in a hospital-based series of patients with gynecological cancer. *Journal of Genetic Counseling,* 22: 345-357.

Demsky, R., McCuaig, J., Maganti, M., Murphy, K. J., Rosen, B. and Armel, S. R. (2013). Keeping it simple: genetics referrals for all invasive serous ovarian cancers. *Gynecologic oncology,* 130: 329-333.

den Heijer M., Seynaeve, C., Vanheusden K., et al. (2013). Long-term psychological distress in women at risk for hereditary breast cancer adhering to regular surveillance: a risk profile. *Psycho-Oncology,* 22: 598-604.

Di Mattei, V., Bernardi, M. and Carnelli, L. (2017). Psychological distress and other aspects regarding cancer genetic testing. In *Counseling and Coaching in Times of Crisis and Transition. From Research to Practice,* edited by L. Nota, and S. Soresi, 206-217. Abingdon, Oxford: Routledge.

Dolbeault, S., Flahault, C., Stoppa-Lyonnet, D. and Brédart, A. (2006). Communication in genetic counseling for breast/ovarian cancer. In *Communication in Cancer Care*, edited by F. Stiefel, 23-36. Springer, Berlin: Heidelberg.

Domchek, S. M., Friebel, T. M., Singer C. F., et al. (2010). Association if risk-reducing surgery in BRCA1 or BRCA2 mutation carriers with cancer risk and mortality. *JAMA,* 304: 967-975.

DudokdeWit, A. C., Tibben, A., Frets, P. G., et al. (1997). BRCA1 in the family: a case description of the psychological implications. *American Journal of Medical Genetics,* 71: 63-71.

Eccles, D. M., Balmana, J., Clune, J., Ehlken, B., et al. (2016). Selecting patients with ovarian cancer for germline BRCA mutation testing: findings from guidelines and a systematic literature review. *Advances in therapy,* 33: 129-150.

Edwards, A., Gray, J., Clarke, A., et al. (2008). Interventions to improve risk communication in clinical genetics: systematic review. *Patient Education and Counselling,* 71: 4-25.

Eijzenga, W., Hahn, D. E., Aaronson, N. K., Kluijt, I. and Bleiker, E. M. (2014). Specific psychosocial issues of individuals undergoing genetic counseling for cancer – a literature review. *Journal of Genetic counseling,* 23: 133-146.

Eijzenga, W., Bleiker, E. M., Hahn, D. E., Kluijt, I., Sidharta, G. N., Gundy, C., and Aaronson, N. K. (2014). Psychosocial aspects of hereditary cancer (PAHC) questionnaire: development and testing of a screening questionnaire for use in clinical cancer genetics. *Psycho-Oncology,* 23: 862-869.

Esplen, M. J., Toner, B., Hunter, J., Glendon, G., Butler, K. and Field, B. (1998). A group therapy approach to facilitate integration of risk information for women at risk for breast cancer. *The Canadian Journal of Psychiatry,* 43: 375-380.

Esplen, M. J. (2006). Psychological aspects of genetic testing for adult-onset hereditary disorders. In *Geneting Tesing: Care, Consent and Liabilit,* edited by N. Sharpe, and R. Carter, 61-77. Hoboken, NJ: John Wiley & Sons Inc.

Esplen, M. J. and Bleiker, E. M. A. (2015). Psychosocial Issues in Genetic Testing for Breast/Ovarian Cancer. In *Psycho-Oncology*, edited by J. C. Holland, W. S. Breitbart, P. N. Butow, P. B. Jacobsen, M. J. Loscalzo, and R. McCorkle, 71-76. New York: Oxford University Press.

Esplen, M. J., Hunter, J., Leszcz, M., et al. (2004). A multicenter study of supportive-expressive group therapy for women with BRCA1/BRCA2 mutations. *Cancer,* 101: 2327-2340.

Esplen, M. J., Cappelli, M., Wong, J., et al. (2011). Development of a Psychosocial Risk Screening Tool for Genetic Testing. Paper presented at the 12th International Meeting on Psychosocial Aspects of Hereditary Cancer, Amsterdam, The Netherlands.

Esplen, M. J., Toner, B., Hunter, J., et al. (2000). A supportive-expressive group intervention for women with a family history of breast cancer: results of a phase II study. *Psycho-Oncology,* 9: 243-252.

Evans, D. G., Baildam, A. D., Anderson, E., et al. (2009). Risk reducing mastectomy: outcomes in 10 European countries. *Journal of Medical Genetics,* 46: 254-258.

Fadaei, S., Janighorban, M., Mehrabi, T., et al. (2011). Effects of cognitive behavioral counseling on body Image following mastectomy. *Journal of Research in Medical Sciences,* 16: 1047-1054.

Farina, S. and Crimi, M. (2017). Genetic counseling: Recommendations for the psychological support within an integrated model. *Journal of Psychology and Cognition,* 2: 198-208.

Finch, A. P., Lubinski, J., Møller, P., et al. (2014). Impact of oophorectomy on cancer incidence and mortality in women with a BRCA1 or BRCA2 mutation. *Journal of Clinical Oncology,* 32: 1547-1553.

Finch, A. and Narod, S. A. (2011). Quality of life and health status after prophylactic salpingo-oophorectomy in women who carry a BRCA mutation: A review. *Maturitas,* 70: 261-265.

Finch, A., Evans, G. and Narod, S. A. (2012). BRCA carriers, prophylactic salpingo-oophorectomy and menopause: clinical management considerations and recommendations. *Womens Health,* 8: 543-555.

Finch, A., Metcalfe, K. A., Chiang, J., et al. (2013). The impact of prophylactic salpingo-oophorectomy on quality of life and psychological distress in women with a BRCA mutation. *Psycho-Oncology,* 22: 212-219.

Fisher, B., Costantino, J. P., Wickerham, D. L., et al. (1998). Tamoxifen for prevention of breast cancer: report of the National Surgical Adjuvant Breast and Bowel Project P-1 Study. *Journal of the National Cancer Institute,* 90: 1371-1388.

Fisher, B., Costantino, J. P., Wickerham, D. L., et al. (2005). Tamoxifen for the prevention of breast cancer: current status of the National Surgical Adjuvant Breast and Bowel Project P-1 study. *Journal of the National Cancer Institute,* 97: 1652-1662.

Fisher, C. L., Maloney, E., Glogowsky, E., Hurley, K., Edgerson, S., Lichtenthal, W. G., Kissane, D. and Bylund, C. (2014). Talking about familial breast cancer risk: topics and strategies to enhance mother-daughter interactions. *Qualitative Health Research,* 24: 517-535.

Forrest, K., Simpson, S. A., Wilson, B. J., van Teijlingen, E. R., McKee, L., Haites, N. and Matthews, E. (2003). To tell or not to tell: barriers and facilitators in family communication about genetic risk. *Clinical Genetics,* 64: 317-326.

Foster, C., Watson, M., Eeles, R., et al. (2007). Predictive genetic testing for BRCA1/2 in a UK clinical cohort: three year follow-up. *British Journal of Cancer,* 96: 718-724.

Friebel, T. M., Domchek, S. M. and Rebbeck, T. R. (2014). Modifiers of cancer risk in BRCA1and BRCA2 Mutation carriers: A systematic Review and Meta-Analysis. *Journal of the National Cancer Institute,* 106: dju091.

Frost, M. H., Hoskin, T. L., Hartmann, L. C., Degnim, A. C., Johnson, J. L. and Boughey, J. C. (2011). Contralateral prophylactic mastectomy: long-term consistency of satisfaction and adverse effects and the significance of informed decision-making, quality of life, and personality traits. *Annals of Surgical Oncology,* 18: 3110.

Gahm, J., Wickman, M. and Brandberg, Y. (2010). Bilateral prophylactic mastectomy in women with inherited risk of breast cancer – prevalence of pain and discomfort, impact on sexuality, quality of life and feelings of regret two years after surgery. *The Breast,* 19: 462-469.

Galvin, K. M and Young, M. A. (2010). Family systems theory and genetics: Synergistic Interconnections. In *Family communication about genetics,* edited by C. Gaff and C.L. Bylund, 102-119. New York: Oxford University Press.

Gavaruzzi, T., Tasso, A., Franiuk, M., Varesco, L. and Lotto, L. (2017). A psychological perspective on factors predicting prophylactic salpiongo-oophorectomy in a sample of Italian women from the general population. Results from a hypothetical study in the context of BRCA mutations. *Journal of Genetic counseling,* 26: 1144-1152.

Geer, K. P., Ropka, M. E., Cohn, W. F., Jones, S. M. and Miesfeldt, S. (2001). Factors influencing patients' decisions to decline cancer genetic counseling services. *Journal of Genetic counseling,* 10: 25-40.

Geiger, A. M., Nekhlyudov, L., Herrinton, J. L., et al. (2007). Quality of life after bilateral prophylactic mastectomy. *Annals of Surgical Oncology,* 14: 686-694.

George, A., Riddell, D., Seal, S., et al. (2016). Implementing rapid, robust, cost-effective, patient-centred, routine genetic testing in ovarian cancer patients. *Scientific reports,* 6: 29506.

George, A., Stan, K. and Susana, B. (2017). Delivering widespread BRCA testing and PARP inhibition to patients with ovarian cancer. *Nature Reviews Clinical Oncology,* 14: 284-296.

Gleeson, M., Meiser, B., Barlow-Stewart, K., et al. (2013). Communication and information needs of women diagnosed with ovarian cancer regarding treatment-focused genetic testing. *Oncology Nursing Forum,* 40: 275-283.

Godard, B., Pratte, A., Dumont, M., Simard-Lebrun, A. and Simard, J. (2007). Factors associated with an individual's decision to withdraw from genetic testing for breast and ovarian cancer susceptibility: implications for counseling. *Genetic Testing,* 11: 45-54.

Gopie, J. P., Mureau, M. A., Seynaeve, C., ter Kuile, M. M., Menke-Pluymers, M. B., Timman, R. and Tibben, A. (2013). Body image issues after bilateral prophylactic mastectomy with breast reconstruction in healthy women at risk for hereditary breast cancer. *Familial Cancer,* 12: 479-487.

Gopie, J. P., Vasen, H. F. and Tibben, A. (2012). Surveillance for hereditary cancer: does the benefit outweight the psychological burden? – a systematic review. *Critical Reviews in Oncology/Hematology,* 83: 329-340.

Goss, P. E., Ingle, J. N., Alés-Martinez, J. E., et al. (2011). Exemestane for breast-cancer prevention in postmenopausal women. *New England Journal of Medicine,* 364: 2381-2391.

Grassi, L. and Riba, M. (2012). *Clinical Psycho-Oncology.* John Wiley & Sons: John-Wiley & Sons, Ltd.

Graves, K. D., Wenzel, L., Schwartz, M. D., et al. (2010). Randomized controlled trial of a psychosocial telephone counseling intervention in BRCA1 and BRCA2 mutation carriers. *Cancer, Epidemiology, Biomarkers & Prevention,* 19: 648-654.

Graves, K. D., Vegella, P., Poggi, E. A., et al. (2012). Long-term psychosocial outcomes of BRCA1/BRCA2 testing: differences across affected status and risk-reducing surgery choice. *Cancer Epidemiology and Prevention Biomarkers,* 21: 445-455.

Halbert, C. H., Stopfer, J. E., McDonald, J., Weathers, B., Collier, A., Troxel, A. B. and Domchek, S. (2011). Long-term reactions to genetic testing for BRCA1 and BRCA2 mutations: does time heal women's concerns? *Journal of Clinical Oncology,* 29: 4302-4306.

Hallowell, N., Baylock, B., Heiniger, L., Butow, P. N., Patel, D., Meiser, B. and Price, M. A. (2012). Looking different, feeling different: women's reactions to risk-reducing breast and ovarian surgery. *Familial Cancer,* 11: 215-224.

Hamilton, J. G., Lobel, M. and Moyer, A. (2009). Emotional distress following genetic testing for hereditary breast and ovarian cancer: a meta-analytic review. *Health Psychology,* 28: 510-518.

Haroun, I., Graham, T., Poll, A., et al. (2011). Reasons for risk- reducing mastectomy versus MRI-screening in a cohort of women at high hereditary risk of breast cancer. *The Breast,* 20: 254-258.

Heemskerk-Gerritsen, B. A., Rookus, M. A., Aalfs, C. M., et al. (2015). Improved overall survival after contralateral risk-reducing mastectomy in BRCA1/2 mutation carriers with a history of unilateral breast cancer: a prospective analysis. *International Journal of Cancer,* 136: 668-677.

Heemskerk-Gerritsen, B. A., Seynaeve, C., van Asperen, C. J., et al. (2015). Breast cancer risk after salpingo-oophorectomy in healthy BRCA1/2 mutation carriers: revisiting the evidence for risk reduction. *Journal of the National Cancer Institute,* 107: djv033.

Heiniger, L., Butow, P. N., Coll, J., et al. (2015). Long-term outcomes of risk-reducing surgery in unaffected women at increased familial risk of breast and/or ovarian cancer. *Familial Cancer,* 14: 105-115.

Hirschberg, A. M., Chan-Smutko, G. and Pirl, W. F. (2015). Psychiatric implications of cancer genetic testing. *Cancer,* 121: 341-360.

Hirschhorn, K., Fleisher, L. D., Godmilow, L., et al. (1999). Duty to re-contact. *Genetics in Medicine,* 1: 171-172.

Ho, S. M., Ho, J. W., Bonanno, G. A., Chu, A. T. and Chan, E. M. (2010). Hopefulness predicts resilience after hereditary colorectal cancer genetic testing: a prospective outcome trajectories study [serial online]. *BMC Cancer,* 10: 279.

Høberg-Vetti, H., Bjorvatn, C., Fiane, B. E., et al. (2016). BRCA1/2 testing in newly diagnosed breast and ovarian cancer patients without prior genetic counseling: the DNA-BONus study. *European Journal of Human Genetics*, 24: 881-888.

Høberg-Vetti, H., Eide, G. E., Siglen, E., Listøl, W., Haavind, M. T., Hoogerbrugge, N. and Bjorvatn, C. (2019). Cancer-related distress in unselected women with newly diagnosed breast or ovarian cancer undergoing BRCA1/2 testing without pretest genetic counseling. *Acta Oncologica,* 58: 175-181.

Hoffman-Andrews, L. (2018). The known unknown: the challenges of genetic variants of uncertain significance in clinical practice. *Journal of Law and the Biosciences,* 4: 648-657.

House of Lords. (2009). *Genomic medicine* (HL Paper 107-I). London: The Stationary Office.

Howard, A. F., Balneaves, L. G., Bottorff, J. L. and Rodney, P. (2011). Preserving the self: the process of decision making about hereditary breast cancer and ovarian cancer risk reduction. *Qualitative Health Research,* 21: 502-519.

Jacobs, C., Patch, C. and Michie, S. (2019). Communication about genetic testing with breast and ovarian cancer patients: a scoping review. *European Journal of Human Genetics,* 27: 511-524.

Jacobs, I. J., Menon, U., Ryan, A., Gentry-Maharaj, A., Burnell, M., Kalsi, J. K. and Skates, S. J. (2016). Ovarian cancer screening and mortality rate in the UK Collaborative Trial of Ovarian Cancer Screening (UKCTOCS): a randomized controlled trial. *The Lancet,* 387: 945-956.

Jernström, H., Lubinski, J., Lynch, H. T., et al. (2004). Breastfeeding and the risk of breast cancer in BRCA1 and BRCA2 mutation carriers. *Journal of the National Cancer Institute,* 96: 1094-1098.

Julian-Reyner, C., Eisinger, F., Chabal, F., et al. (2000). Disclosure to the family of breast/ovarian cancer genetic test results: patients' willingness and associated factors. *American Journal of Medical Genetics,* 94: 13-18.

Julian-Reynier, C., Welkenhuysen, M., Hagoel, L., Decruyenaere, M. and Hopwood, P. (2003). Risk communication strategies: state of the art and effectiveness in the context of cancer genetic services. *European Journal of Human Genetics,* 11: 725-736.

Kasparian, N. A., Wakefield, C. E. and Meiser, B. (2007). Assessment of psychosocial outcomes in genetic counseling research: an overview of available measurement scales. *Journal of Genetic counseling,* 16: 693-712.

Kathawala, R. J., Kudelka, A. and Rigas, B. (2018). The chemoprevention of ovarian cancer: the need and the options. *Current Pharmacology Reports,* 4: 250-260.

Kenen, R., Arden-Jones, A. and Eeles, R. (2004). We are talking, but are they listening? Communication patterns in families with a history of breast/ovarian cancer (HBOC). *Psycho-Oncology,* 13: 335-345.

Keogh, L. A., van Vliet, C. M., Studdert, D. M., et al. (2009). Is the uptake of genetic testing for colorectal cancer influenced by knowledge of insurance implications? *Medical Journal of Australia,* 191: 255-258.

King, M. C., Marks, J. H. and Mandell, J. B. (2003). Breast and Ovarian Cancer Risks Due to Inherited Mutations in BRCA1 and BRCA2. *Science,* 302: 643-646.

Klitzman, R. (2012). *Am I my Genes? Confronting Fate & Familiy Secrets in the Age of Gentic Testing.* New-York: Oxford University Press.

Kotsopoulos, J., Lubinski, J., Salmena, L., et al. (2012). Breastfeeding and the risk of breast cancer in BRCA1 and BRCA2 mutation carriers. *Breast Cancer Research,* 14: R42.

Kurian, A. W., Li, Y., Hamilton, A. S., et al. (2017). Gaps in incorporating germline genetic testing into treatment decision-making for early-stage breast cancer. *Clinical Oncology,* 35: 2232-2239.

Lerman, C., Lustbader, E., Rimer, B., et al. (1995). Effects of individualized breast cancer risk counseling: a randomized trial. *Journal of the National Cancer Institute*, 87: 286-292.

Lindor, N. M., McMaster, M. L., Lindor, C. J. and Greene, M. H. (2008). Concise handbook of familial cancer susceptibility syndromes. *JNCI monographs,* 2008: 3-93.

Lipkus, I. M., Crawford, Y., Fenn, K., Biradavolu, M., Binder, R. A., Marcus, A. and Mason, M. (1999). Testing different formats for communicating colorectal cancer risk. *Journal of Health Communication,* 4: 311-324.

Lipkus, I. M. and Hollands, J. G. (1999). The visual communication of risk. *Journal of the National Cancer Institute Monographs,* 25: 149-163.

Lobb, E. A., Butow, P. N. and Meiser, B. (2003). Women's preferences and consultants' risk communication in familial breast cancer consultations: impact on patient outcomes. *Journal of Medical Genetics,* 40: e56.

Lodder, L., Frets, P. G., Trijsburg, R. W., et al. (2001). Psychological impact of receiving a BRCA1/BRCA2 test result. *American Journal of Medical Genetics,* 98: 15-24.

Lostumbo, L., Carbine, N. E. and Wallace, J. (2010). Prophylactic mastectomy for the prevention of breast cancer. *Cochrane Database Systematic Review,* 11: CD002748.

MacDonald, D. J., Sarna, L., van Servellen, G., Bastani, R., Giger, J. N., and Weitzel, J. N. (2007). Selection of family members for communication of cancer risk and barriers to this communication before and after genetic cancer risk assessment. *Genetics in Medicine,* 9: 275-282.

Macrae, L., Navarro de Souza, A., Loiselle, C. G., and Wong, N. (2013). Experience of BRCA1/2 mutation-negative young women from families with hereditary breast and ovarian cancer: a qualitative study. *Hereditary Cancer in Clinical Practice,* 11: 14.

Madalinska, J. B., Hollenstein, J., Bleiker, E., et al. (2005). Quality-of-life effects of prophylactic salpingo-oophorectomy versus gynecologic screening among women at increased risk of hereditary ovarian cancer. *Journal of Clinical Oncology,* 23: 6890-6898.

Madalinska, J. B., van Beurden, M., Bleiker, E. M., et al. (2006). The impact of hormone replacement therapy on menopausal symptoms in younger high-risk women after prophylactic salpingo-oophorectomy. *Journal of Clinical Oncology,* 24: 3576-3582.

Maeland, M. K., Eriksen, E. O. and Synnes, O. (2014). The loss of a mother and dealing with genetic cancer risk: women who have undergone prophylactic removal of the ovaries. *European Journal of Oncology Nursing,* 18: 521-526.

Maheu, C. and Thorne, S. (2008). Receiving inconclusive genetic test results: an interpretative description of the BRCA1/2 experience. *Research in Nursing & Health,* 31: 553-562.

Marchetti, C., De Felice, F., Palaia, I., et al. (2014). Risk-reducing salpingo-oophorectomy: a meta-analysis on impact on ovarian cancer risk and all cause mortality in BRCA 1 and BRCA 2 mutation carriers. *BMC Women Health,* 14: 150.

Maunder, R. G., and Hunter, J. J. (2008). Attachment relationships as determinants of physical health. *The Journal of the American Academy of Psychoanalysis and Dynamic Psychiatry,* 36: 11-32.

Mayo Foundation for Medical Education and Research. (2019). BRCA gene test for breast and ovarian cancer risk. Accessed August 1. https://www.mayoclinic.org/tests-procedures/brca-gene-test/about/pac-20384815?p=1.

McGivern, B., Everett, J., Yager, G. G., Baumiller, R. C., Hafertepen, A. and Saal, H. M. (2004). Family communication about positive BRCA1 and BRCA2 genetic test results. *Genetics in Medicine,* 6: 503-509.

Meadows, R., Padamsee, T. J., and Paskett, E. D. (2018). Distinctive psychological and social experiences of women choosing prophylactic oophorectomy for cancer prevention. *Health Care for Women International,* 39: 595-616.

Meiser, B. (2005). Psychological impact of genetic testing for cancer susceptibility: an update of the literature. *Psycho-Oncology,* 14: 1060-1074.

Meiser, B. (2012). Getting to the point: what women newly diagnosed with breast cancer want to know about treatment-focused genetic testing. *Oncology Nursing Forum,* 39: E101.

Meiser, B., Gleeson, M., Kasparian, N., et al. (2012). There is no decision to make: experiences and attitudes toward treatment-focused genetic testing among women diagnosed with ovarian cancer. *Gynecologic oncology*, 124: 153-157.

Meiser, B., Butow, P., Friedlander, M., et al. (2002). Psychological impact of genetic testing in women from high-risk breast cancer families. *European Journal of Cancer*, 38: 2025-2031.

Metcalfe, K. A., Finch, A., Poll, A., Horsman, D., Kim-Sing, C., Scott, J., Royer, R., Sun, P. and Narod, S. A. (2009). Breast cancer risks in women with a family history of breast or ovarian cancer who have tested negative for a BRCA1 or BRCA2 mutation. *British Journal of Cancer,* 100: 421-425.

Metcalfe, K. A., Birenbaum-Carmeli, D., Lubinski, J., et al. (2008). Hereditary Breast Cancer Clinical Study Group. International variation in rates of uptake of preventive options in BRCA1 and BRCA2 mutation carriers. *International Journal of Cancer*, 122: 2017-2022.

Metcalfe, K. A., Esplen, M. J., Goel, V. and Narod, S. A. (2004). Psychosocial functioning in women who have undergone bilateral prophylactic mastectomy. *Psycho-Oncology,* 13: 14-25.

Metcalfe, K. A., Esplen, M. J., Goel, V. and Narod, S. A. (2005). Predictors of quality of life in women with a bilateral prophylactic mastectomy. *The Breast Journal,* 11: 65-69.

Michelsen, T. M., Dørum, A. and Dahl, A. A. (2009). A controlled study of mental distress and somatic complaints after risk-reducing salpingo-oophorectomy in women at risk for hereditary breast ovarian cancer. *Gynecologic Oncology,* 113: 128-133.

Miller, S. M., Roussi, P., Daly, M. B., et al. (2005). Enhanced counseling for women undergoing BRCA1/2 testing: impact on subsequent decision making about risk reduction behaviors. *Health Education & Behavior,* 32: 654-667.

Moldovan, R., Keating, S. and Clancy, T. (2015). The impact of risk-reducing gynaecological surgery in premenopausal women at high risk of endometrial and ovarian cancer due to Lynch syndrome. *Familial Cancer,* 14: 51-60.

Møller, P., Hagen, A. I., Apold, J., et al. (2007). Genetic epidemiology of BRCA mutations–family history detects less than 50% of the mutation carriers. *European Journal of Cancer,* 43: 1713-1717.

Murakami, Y., Okamura, H., Sugano, K., et al. (2004). Psychologic distress after disclosure of genetic test results regarding hereditary nonpolyposis colorectal carcinoma. *Cancer,* 101: 395-403.

Murray, M. L., Cerrato, F., Bennett, R. L. and Jarvik, G. P. (2011). Follow-up of carriers of BRCA1 and BRCA2 variants of unknown significance: variant reclassification and surgical decisions. *Genetics in Medicine,* 12: 998-1005.

National Cancer Institute. (2019). Breast Cancer Risk Assessment Tool. Accessed July 24. https://bcrisktool.cancer.gov/.

National Comprehensive Cancer Network. (2016). *NCCN clinical practice guidelines in oncology: genetic/familial high-risk assessment: breast and ovarian.* Fort Washington: National Comprehensive Cancer Network, 1-105.

National Institute for Health and Care Excellence guidelines. (2013). Familial breast cancer: classification and care of people at risk of familial breast cancer and management of breast cancer and related risks in people with a family history of breast cancer. Accessed August 1. http://www.nice.org.uk/guidance/cg164.

Nycum, G., Avard, D. and Knoppers, B. M. (2009). Factors influencing intrafamilial communication of hereditary breast and ovarian cancer genetic information. *European Journal of Human Genetics,* 17: 872-880.

O'Neill, S. C., Rini, C., Goldsmith, R. E., Valdimarsdottir, H., Cohen, L. H. and Schwartz, M. D. (2009). Distress among women receiving uninformative BRCA1/2 results: 12-month outcomes. *Psycho-Oncology: Journal of the Psychological, Social and Behavioral Dimensions of Cancer,* 18: 1088-1096.

O'Neill, S. C, DeMarco, T., Peshkin, B. N., et al. (2006). Tolerance for uncertainty and perceived risk among women receiving uninformative *BRCA1/2* test results. *American Journal of Medical Genetics Part C: Seminars in Medical Genetics,*142C: 251-259.

Okabayashi, H. (2017). The Formation of Mutual Understanding in Conversation: An Embodied Approach. *International Journal of Psychological and Behavioral Sciences*, 11: 563-569.

Paluch-Shimon, S., Cardoso, F., Sessa, C., Balmana, J., Cardoso, M. J., Gilbert, F. and Senkus, E. (2016). Prevention and screening in BRCA mutation carriers and other breast/ovarian hereditary cancer syndromes: ESMO Clinical Practice Guidelines for cancer prevention and screening. *Annals of Oncology,* 27(suppl_5): v103-v110.

Patenaude, A. F., and Katherine, A. S. (2017). Issues arising in psychological consultations to help parents talk to minor and young adult children about their cancer genetic test result: a guide to providers. *Journal of Genetic counseling,* 26: 251-260.

Patenaude, A. F., Dorval, M., DiGianni, L. S., Schneider, K. A., Chittenden, A. and Garber, J. E. (2006). Sharing BRCA1/2 test results with first-degree relatives: factors predicting who women tell. *Journal of Clinical Oncology,* 24: 700-706.

Peterson, S. K. (2005). The role of family in genetic testing: theoretical perspectives, current knowledge, and future directions. *Health Education & Behavior,* 32: 627-639.

Petzel, S. V., Vogel, R. I., Bensend, T., Leininger, A., Argenta, P. A. and Geller, M. A. (2013). Genetic risk assessment for women with epithelial ovarian cancer: referral patterns and outcomes in a university gynecologic oncology clinic. *Journal of Genetic counseling,* 22: 662-673.

Phelps, C., Bennett, P., Jones, H., Hood, K., Brain, K. and Murray, A. (2010). The development of a cancer genetic-specific measure of coping: the GRACE. *Psycho-Oncology,* 19: 847-854.

Phelps, C., Bennett, P., Iredale, R., Anstey, S. and Gray, J. (2006). The development of a distraction-based coping intervention for women waiting for genetic risk information: a phase 1 qualitative study. *Psycho-Oncology: Journal of the Psychological, Social and Behavioral Dimensions of Cancer*, 15: 169-173.

Pieterse, A. H., Van Dulmen, A. M., Beemer, F. A., Bensing, J. M. and Ausems, M. G. (2007). Cancer genetic counseling: communication and counselees' post-visit satisfaction, cognitions, anxiety, and needs fulfillment. *Journal of Genetic Counseling,* 16: 85-96.

Pieterse, A., van Dulmen, S., Ausems, M., Schoemaker, A., Beemer, F. and Bensing, J. (2005). QUOTE-gene(ca): development of a counselee-centered instrument to measure needs and preferences in genetic counseling for hereditary cancer. *Psycho-Oncology,* 14: 361-375.

Pieterse, K., van Dooren, S., Seynaeve, C., et al. (2007). Passive coping and psychological distress in women adhering to regular breast cancer surveillance. *Psycho-Oncology*, 6:851-858.

Plaskocinska, I., Shipman, H., Drummond, J., et al. (2016). New paradigms for BRCA1/BRCA2 testing in women with ovarian cancer: results of the Genetic Testing in Epithelial Ovarian Cancer (GTEOC) study. *Journal of Medical Genetics,* 53: 655-661.

Pruthi, S., Heisey, R. E. and Bevers, T. B. (2015). Chemoprevention for breast cancer. *Annals of Surgical Oncology,* 22: 3230-3235.

Read, C. Y., Perry, D. J. and Duffy, M. E. (2005). Design and psychometric evaluation of the Psychological Adaptation to Genetic Information Scale. *Journal of Nursing Scholarship*, 37: 203-208.

Reichelt, J. G., Heimdal, K., Moller, P. and Dahl, A. A. (2004). BRCA1 testing with definitive results: a prospective study of psychological distress in a large clinic-based sample. *Familial Cancer,* 3: 21-28.

Reiff, M., Ross, K., Mulchandani, S., Propert, K. J., Pyeritz, R. E., Spinner, N. B., and Bernhardt, B. A. (2013). Physicians' perspectives on the uncertainties and implications of chromosomal microarray testing of children and families. *Clinical Genetics,* 83: 23-30.

Resta, R., Biesecker, B. B., Bennett, R. L., Blum, S., Estabrooks Hahn, S., Strecker, M. N. and Williams, J. L. (2006). A new definition of genetic counseling: National Society of Genetic Counselors' Task Force report. *Journal of Genetic Counseling,* 15: 77-83.

Richards, S., Aziz, N., Bale, S., et al. (2015). Standards and Guidelines for the Interpretation of Sequence Variants: A Joint Consensus Recommendation of the American College of Medical Genetics and Genomics and the Association for Molecular Pathology. *Genetics in Medicine,* 17: 405-423.

Richter, S., Haroun, I., Graham, T. C., Eisen, A., Kiss, A. and Warner, E. (2013). Variants of unknown significance in BRCA testing: impact on risk perception, worry, prevention and counseling. *Annuals of Oncology*, 24: viii69-viii74.

Ridge, Y., Panabaker, K., McCullum, M., Portigal-Todd, C., Scott, J. and McGillivray, B. (2009). Evaluation of group genetic counseling for hereditary breast and ovarian cancer. *Journal of genetic counseling*, 18: 87-100.

Roa, B. B., Boyd, A. A., Volcik, K. and Richards, C. S. (1996). Askenazi Jewish population frequencies for common mutations in BRCA1 and BRCA2. *Nature Genetics,* 14: 185-187.

Rodin, G., Mackay, J. A., Zimmermann, C., et al. (2009). Clinician–patient communication: a systematic review. *Supportive Care in Cancer*, 17: 627-644.

Ropka, M. E., Wenzel, J., Phillips, E. K., Siadaty, M. and Philbrick, J. T. (2006). Uptake rates for breast cancer genetic testing: a systematic review. *Cancer Epidemiology, Biomarkers & Prevention*, 15: 840-855.

Rothman, A. J., and Kiviniemi, M. T. (1999). Treating people with information: an analysis and review of approaches to communicating health risk information. *Journal of the National Cancer Institute Monographs*, 25: 44-51.

Rothwell, E., Kohlmann, W., Jasperson, K., Gammon, A., Wong, B. and Kinney, A. (2012). Patient outcomes associated with group and individual genetic counseling formats. *Familial Cancer,* 11: 97-106.

Sahin, I., Isik, S., Alhan, D., Yildiz, R., Aykan, A. and Ozturk, E. (2013). One-staged silicone implant breast reconstruction following bilateral nipple-sparing prophylactic mastectomy in patients at high-risk for breast cancer. *Aesthetic Plastic Surgery,* 3: 303-311.

Sarangi, S. (2000). Activity types, discourse types and interactional hybridity: the case of genetic counseling. In *Discourse and social life*, edited by S. Sarangi and M. Coulthard, 1-27. Harlow, Essex: Longman.

Schapira, M., Nattinger, A. and McHorney, C. (2001). Frequency or probability? A qualitative study of risk communication formats used in health care. *Medical Decision Making,* 21: 459-467.

Scherr, C. L., Vasquez, E., Quinn, G. P. and Vadaparampil, S. T. (2014). Genetic counseling for hereditary breast and ovarian cancer among Puerto Rican women living in the United States. *Reviews on recent clinical trials,* 9: 245-253.

Schlich-Bakker, K. J., ten Kroode, H. F., Wárlám-Rodenhuis, C. C., van den Bout, V. and Ausems, M. G. (2007). Barriers to participating in genetic counseling and BRCA testing during primary treatment for breast cancer. *Genetics in Medicine,* 9: 766-777.

Schlich-Bakker, K. J., Ausems, M. G., Schipper, M., ten Kroode, H. F., Wárlám-Rodenhuis, C. C. and van den Bout, J. (2008). BRCA1/2 mutation testing in breast cancer patients: a prospective study of the long-term psychological impact of approach during adjuvant radiotherapy. *Breast Cancer Research and Treatment,* 109: 507-514.

Schroeder, D., Duggleby, W. and Cameron, B. L. (2017). Moving in and out the what-ifs. The experiences of unaffected women linving in families where a breast-cancer 1 or 2 genetic mutation was not found. *Cancer Nursing*, 40: 386-393.

Schwartz, M. D., Isaacs, C., Graves, K. D., et al. (2012). Long-term outcomes of BRCA1/BRCA2 testing: risk reduction and surveillance. *Cancer,* 118: 510-517.

Schwartz, M., Peshkin, B., Hughs, C., Main, D., Isaacs, C. and Lerman, C. (2002). Impact of BRCA1/BRCA2 mutation testing on psychological distress in a clinic-based sample. *Journal of Clinical Oncology,* 20: 514-520.

Sermijn, E., Goelen, G., Teugels, E., Kaufman, L., Bonduelle, M., Neyns, B., and De Greve, J. (2004). The impact of proband mediated information dissemination in families with a BRCA1/2 gene mutation. *Journal of Medical Genetics*, 41: e23.

Shigehiro, M., Kita, M., Takeuchi, S., Ashihara, Y., Arai, M. and Okamura, H. (2015). Study on the psychosocial aspects of risk-reducing salpingo-oophorectomy (RRSO) in BRCA1/2 mutation carriers in Japan: a preliminary report. *Japanese Journal of Clinical Oncology*, 46: 254-259.

Shiloh, S., Koehly, L., Jenkins, J., Martin, J. and Hadley, D. (2008). Monitoring coping style moderates emotional reactions to genetic testing for hereditary nonpolyposis colorectal cancer: a longitudinal study. *Psycho-Oncology*, 17:746-775.

Shipman, H., Flynn, S., MacDonald-Smith, C. F., Brenton, J., et al. (2017). Universal BRCA1/BRCA2 testing for ovarian cancer patients is welcomed, but with care: how women and staff contextualize experiences of expanded access. *Journal of Genetic Counseling,* 26: 1280-1291.

Soran, A., Ibrahim, A., Kanbour, M., et al. (2015). Decision making factors influencing long-term satisfaction with prophylactic mastectomy in women with breast cancer. *American Journal of Clinical Oncology,* 38: 179-183.

Spiegel, D. and Spira, J. (1991). *Supportive-expressive Group Therapy: A Treatment Manual of Psychosocial Intervention for Women with Recurrent Breast Cancer.* Psychosocial Treatment Laboratory, Stanford University School of Medicine, Stanford CA.

Stoffel, E. M., Ford, B., Mercado, R. C., et al. (2008). Sharing genetic test results in Lynch syndrome: communication with close and distant relatives. *Clinical Gastroenterology and Hepatology* 6: 503-509.

Telli, M. L., Jensen, K. C., Vinayak, S., Kurian, A., et al. (2015). Phase II study of gemcitabine, carboplatin, and iniparib as neoadjuvant therapy for triple-negative and BRCA1/2 mutation-associated breast cancer with assessment of a tumore-based measure of genomic instability: PrECOG 0105. *Journal of Clinical Oncology,* 33: 1895-1901.

The US department of Health and Human Services. (2014). Accessed July 30. https://www.fda.gov/drugs.

Trainer, A. H., Lewis, C. R., Tucker, K., Meiser, B., Friedlander, M. and Ward, R. L. (2010). The role of BRCA mutation testing in determining breast cancer therapy. *Nature Reviews Clinical Oncology,* 7: 108-117.

Trepanier, A. M., and Allain, D. C. (2014). Models of service delivery for cancer genetic risk assessment and counseling. *Journal of Genetic counseling,* 23: 239-253.

Tucker, P. E., and Cohen, P. A. (2017). Sexuality and risk-reducing salpingo-oophorectomy. *International Journal of Gynecological Cancer,* 27: 847-852.

Van Dijk, S., Otten, W., Tollenaar, R. A. E. M., Van Asperen, C. J. and Tibben, A. (2008). Putting it all behind: long-term psychological impact of an inconclusive DNA test result for breast cancer. *Genetics in Medicine* 10: 745-750.

van Oostrom, I., Meijers-Heijboer, H., Dulvenvoorden H. J., et al. (2006). Experience of parental cancer in childhood is a risk factor for psychological distress during genetic cancer susceptibility testing. *Annals of Oncology* 17: 1090-1095.

van Oostrom, I., Meijers-Heijboer, H., Dulvenvoorden, H. J., et al. (2007). Prognostic factors for hereditary cancer distress six months after

BRCA1/2 or HNPCC genetic susceptibility testing. *European Journal of Cancer,* 43: 71-77.

van Oostrom, I., Meijers-Heijboer, H., Lodder, L. N., et al. (2003). Long-term psychological impact of carrying a BRCA1/2 mutation and prophylactic surgery: a 5-year follow-up study. *Journal of Clinical Oncology,* 21: 3867-3874.

van Roosmalen, M. S., Stalmeier, P. F., Verhoef, L. C., et al. (2004). Impact of BRCA1/2 testing and disclosure of a positive result on women affected and unaffected with breast or ovarian cancer. *American Journal of Medical Genetics Part A,* 124: 346-355.

Vogel, V. G., Costantino, J. P., Wickerham, D. L., et al. (2010). Update of the National Surgical Adjuvant Breast and Bowel Project Study of Tamoxifen and Raloxifene (STAR) P-2 Trial: Preventing breast cancer. *Cancer Prevention Research,* 3: 696-706.

Vogel, V. G., Costantino, J. P., Wickerham, D. L., et al. (2006). Effects of tamoxifen vs raloxifene on the risk of developing invasive breast cancer and other disease outcomes: the NSABP Study of Tamoxifen and Raloxifene (STAR) P-2 trial. *JAMA,* 295: 2727-2741.

Voorwindem, J. S. and Jaspers, J. P. C. (2016). Prognostic factors for distress after genetic testing for hereditary cancer. *Journal of Genetic counseling,* 25: 495-503.

Vos, J., E. Gómez-García, J. C. Oosterwijk, et al. 2012. Opening the psychological black box in genetic counseling. The psychological impact of DNA testing is predicted by the counselees' perception, the medical impact by the pathogenic or uninformative BRCA1/2-result. *Psycho-Oncology* 21:29-42.

Vos, J., Oosterwijk, J. C., Gomez-Garcia, E., et al. (2012). Exploring the short-term impact of DNA-testing in breast cancer patients: the counselees' perception matters, but the actual BRCA1/2 result does not. *Patient education and counselling,* 86: 239-251.

Vos, J., Otten, W., van Asperen, C., Jansen, A., Menko, F. and Tibben, A. (2008). The counselees' view of an unclassified variant in BRCA1/2: recall, interpretation and impact on life. *Psycho-Oncology,* 17: 822-830.

Wagner Costalas, J., Itzen, M., Malick, J., Babb, J. S., Bove, B., Godwin, A. K., and Daly, M. B. (2003). Communication of BRCA1 and BRCA2 results to at-risk relatives: a cancer risk assessment program's experience. *American Journal of Medical Genetics. Part C: Seminars in Medical Genetics,* 119C: 11-18.

Wakefield, C. E., Meiser, B., Homewood, J., et al. (2007). Development and pilot testing of two decision aids for individuals considering genetic testing for cancer risk. *Journal of Genetic counseling,* 16: 325-399.

Wakefield, C. E., Ratnayake, P., Meiser, B., Suthers, G., Price, M. A., Duffy, J. and Tucker, K. (2011). "For all my family's sake, I should go and find out": an Australian report on genetic counseling and testing uptake in individuals at high risk of breast and/or ovarian cancer. *Genetic Testing and Molecular Biomarkers,* 15: 379-385.

Wasteson, E., Sandelin, K., Brandberg, Y., Wickman, M., and Arver, B. (2011). High satisfaction rate ten years after bilateral prophylactic mastectomy - a longitudinal study. *European Journal of Cancer Care,* 20: 508-513.

Watson, E. K., Henderson, B. J., Brett, J., Bankhead, C. and Austoker, J. (2005). The psychological impact of mammographic screening on women with a family history of breast cancer – a systematic review. *Psycho-Oncology,* 14: 939-948.

Watson, M., Foster, C., Eeles R., et al. (2004). Psychosocial impact of breast/ovarian (BRCA1/2) cancer-predictive genetic testing in a UK multi-centre clinical cohort. *British Journal of Cancer,* 91: 1787-1794.

Weinstein, N. D. (1999). What does it mean to understand a risk? Evaluating risk comprehension. *Journal of the National Cancer Institute Monographs*, 25: 15-20.

Weitzel, J. N., McCaffrey, S. M., Nedelcu, R., MacDonald, D. J., Blazer, K. R., and Culliane, C. A. (2003). Effect of genetic cancer risk assessment in surgical decisions at breast cancer diagnosis. *Archives of Surgery,* 138: 1323-1328.

Wellisch, D. K. and Lindberg, N. M. (2001). A psychological profile of depressed and nondepressed women at high risk for breast cancer. *Psychosomatics,* 42: 330-336.

Wellisch, D. K., Hoffman, A., Goldman, S., Hammerstein, J., Klein, K. and Bell, M. (1999). Depression and anxiety symptoms in women at high risk for breast cancer: pilot study of a group intervention. *American Journal of Psychiatry,* 156: 1644-1645.

Wellisch, D. K., Gritz, E. R., Schain, W., et al. (1991). Psychological functioning of daughters of breast cancer patients. Part I: Daughters and comparison subjects. *Psychosomatics* 32: 324-336.

Wevers, M. R., Hahn, D. E., Verhoef, S., et al. (2012). Breast cancer genetic counseling after diagnosis but before treatment: a pilot study on treatment consequences and psychological impact. *Patient Education and Counseling,* 89: 89-95.

Wevers, M. R., Aaronson, N. K., Bleiker, E., et al. (2017). Rapid genetic counseling and testing in newly diagnosed breast cancer: patients' and health professionals' attitudes, experiences and evaluation of effects on treatment decision-making. *Journal of Surgical Oncology,* 116: 1029-1039.

Willis, A. M, Smith, S. K., Meiser, B.,. Ballinger, M. L., Thomas, D. M., and Young, M.-A. (2017). Sociodemographic, psychosocial and clinical factors associated with uptake of cancer genetic counseling for hereditary cancer: a systematic review. *Clinical Genetics,* 92: 121-133.

Wilson, B. J., Forrest, K., van Teijlingen, E. R., McKee, L., Haites, N., Matthews, E. and Simpson, S. A. (2004). Family communication about genetic risk: the little that is known. *Public Health Genomics*, 7: 15-24.

Wong, M., Ratner, J., Gladstone, K. A., Davtyanm, A. and Koopman, C. (2010). Children's perceived social support after a parent is diagnosed with cancer. *Journal of Clinical Psychology in Medical Settings,* 17: 77-86.

Wright, S., Porteous, M., Stirling, D., Lawton, J., Young, O., Goutley, C. and Hallowell, N. (2018). Patients' views of treatment-focused genetic testing (TFGT): some lessons for the mainstreaming of BRCA1 and BRCA2 testing. *Journal of Genetic Counseling,* 27: 1459-1472.

Yoon, S. Y., Thong, M. K., Taib, N. A., Yip, C. H. and Teo, S. H. (2011). Genetic counseling for patients and families with hereditary breast and ovarian cancer in a developing Asian country: an observational descriptive study. *Familial Cancer*, 10: 199-205.

Zhang, S., Royer, R., Li, S., McLaughlin, J. R., et al. (2011). Frequencies of BRCA1 and BRCA2 mutations among 1,342 unselected patients with invasive ovarian cancer. *Gynecologic Oncology*, 121: 353-357.

Zilliacus, E., Meiser, B., Gleeson, M., Watts, K., Tucker, K., Lobb, E. A. and Mitchell, G. (2012). Are we being overly cautious? A qualitative inquiry into the experiences and perceptions of treatment-focused germline BRCA genetic testing amongst women recently diagnosed with breast cancer. *Supportive Care in Cancer*, 20: 2949-2958.

In: A Comprehensive Guide …
Editor: Benjamin A. Kepert

ISBN: 978-1-53616-975-1
© 2020 Nova Science Publishers, Inc.

*Chapter 2*

# NEXT-GENERATION SEQUENCING EXPERIENCE: IMPACT OF EARLY DIAGNOSIS OF USHER SYNDROME

## *Caitlin Wright\*, Aimee Brown, Christina Hurst, Amrita Mukherjee, Michael Albert, Jr., MD, Gerald McGwin, PhD and Nathaniel H. Robin, MD*

University of Alabama at Birmingham, Birmingham, Alabama, US

## ABSTRACT

*Purpose*: The aim of this study was to assess the parental psychosocial implications, such as emotions and coping, of earlier diagnosis of Usher syndrome via genetic testing compared to parents of children who were diagnosed later via ophthalmologic findings.

*Method*: Thirty-six participants were recruited through an online posting on the Usher Syndrome Coalition website. Two comparison groups were formed based on the method of diagnosis (i.e., genetic diagnosis vs. ophthalmologic diagnosis). Semi-structured interviews were recorded and transcribed. Comparison, using thematic and statistical

---

\* Corresponding Author's E-mail: cwright1@uab.edu.

analysis, of psychosocial impact on parents of children diagnosed early (via genetic testing) and later (based on ophthalmologic findings) was completed.

*Results*: There were no statistically significant differences in emotions between the two groups of participants, suggesting that earlier diagnosis via genetic testing does not lead to increased anxiety or psychosocial issues for parents. Additional themes identified from parent interviews and their application to patient care are described.

*Conclusion*: Earlier diagnosis of Usher syndrome via genetic testing does not cause a more harmful emotional impact than later diagnosis via ophthalmologic findings. In fact, there are multiple benefits to earlier diagnosis via genetic testing. Earlier diagnosis allows parents to emotionally process and prepare the child for independence throughout life.

# INTRODUCTION

Usher syndrome (USH) is an autosomal recessive disorder characterized by congenital or early-onset sensorineural hearing loss (SNHL) and later onset progressive vision loss due to retinitis pigmentosa (RP). There are three main subtypes, differentiated based on the presence of vestibular dysfunction and age of onset of RP (Lentz and Keats 1999). Management guidelines have been developed for USH, but timing of evaluations and treatments can vary depending on type and age of onset.

Since the vision loss associated with Usher syndrome has varying ages of onset, it is important that all affected individuals have routine ophthalmologic evaluations with dilation to detect onset of RP, as well as potentially treatable conditions, like cataracts. Currently, there are not any useful prevention or treatment options for RP, but people with RP might benefit from early use of visual aids and UV-A and UV-B blocking sunglasses (Fahim, Daiger, and Weleber 2000).

Regarding hearing loss, individuals with USH Type 1 are recommended to consider cochlear implantation (CI). A CI at a younger age correlates with improved chances for speech development (Lentz and Keats 1999). Individuals with USH Type 2 benefit from hearing aids that can help to normalize speech. Individuals with Types 2 or 3 should have routine audiologic evaluations in order to detect any changes in hearing ability. Later in life, individuals with Type 2 or 3 might consider CI if their hearing loss has progressed (due to environmental factors or susceptibility to presbycusis). It can also be helpful for the individual's family to learn forms of communication other than speech, like sign language, or even tactile sign language (Lentz and Keats 1999). Learning tactile sign language can be especially helpful since RP leads to vision loss later in life.

Supervised sporting events or physical therapy are helpful with the vestibular dysfunction that individuals who have USH Type 1 or 3 experience. Physical activity and motor coordination can be helpful in helping individuals with balance issues.

Historically, for patients without a family history, USH has been diagnosed when an adolescent or adult with SNHL manifests problems in vision, which are subsequently diagnosed as RP. Therefore, it is common for an individual who may have thought they simply have SNHL, to be devastated by the realization of the pending loss of another sense. Many individuals, for example, use sign language as their means of communication, which will no longer be possible with the progression of RP, threatening their independence. (Ellis and Hodges, 2013; Brabyn, Haegerstrom-Portnoy, and Lott 2007; Zannon de Andrade Figueiredo et al. 2013). A common sentiment among individuals with USH is that they wish they had been able to prepare for vision loss by learning new forms of communication that could be used once they were unable to sign (Zannon de Andrade Figueiredo et al. 2013; Damen, Krabbe, Kilsby, and Mylanus 2005).

## Table 1. Potential benefits of earlier diagnosis

| |
|---|
| Relief of parental guilt |
| Additional time to emotionally process diagnosis before helping child process diagnosis |
| Access to additional services (e.g., bilateral CI) |
| Ability to participate in research before onset of vision loss |
| Additional time to prepare for the child's future |
| Ability to learn various ways to cope |

The diagnosis of USH also has a significant impact on the parents of an affected child. Most (95%) of children with SNHL are born to hearing parents (Mitchell and Karchmer 2004). There are many stages of emotions surrounding a diagnosis of hearing loss, including fear, anger, and confusion. (Luterman 1999). Should they utilize sign language? Should their children undergo Cochlear implantation (CI)? How do they access the best educational resources? These decisions are difficult with a child with SNHL, but are made far more complicated with the prospective of progressive vision loss (DeCarlo, McGwin, Bixler, Wallander, and Owsley 2012). In addition, there is the stress related to the anticipation of additional vision loss. It would seem that the timing of the diagnosis of USH would have an impact on these psychosocial issues.

There has been significant research into the genetic basis of SNHL (Idan, Shivatzki, and Avraham 2013). Over a hundred genes have been discovered, with more to come. However, until relatively recently, genetic testing was limited to testing one gene at a time. This approach was inefficient and limited, as only a few genes (e.g., *GJB2*, *SLC26A7*) could be tested for the given patient. This changed with the advent of Next-Generation DNA sequencing (NGS). This technology allowed for testing of dozens to hundreds of genes simultaneously and relatively inexpensively. NGS type testing is ideal for SNHL, as most genes are associated with presentation of only SNHL and no other distinguishing features. NGS has led to a much higher rate of genetic diagnosis in children with SNHL (Vona, et al. 2014). However, it has also led to much earlier diagnosis of USH than what was typical in the pre-NGS era, when USH was diagnosed with the onset of vision symptoms in older individuals with hearing loss. Diagnosis of USH before the onset of RP, which is in stark contrast to the age of diagnosis made when using clinical evaluations, may have many benefits and consequences to patients and their families.

While earlier diagnosis provides many potential benefits (See Table 1), it also has the risk of causing harm. Consider that parents have just learned the unexpected news that their child did not pass the newborn hearing screen. Over the next weeks and months, parents come to cope with the realization of having a child with hearing loss, and then they receive a second unpredicted revelation about their child's future: they will lose their vision. Does the benefit of early diagnosis outweigh the psychosocial impact?

In an effort to address this question, we completed this study comparing the experience of parents of children with USH who were diagnosed as young children through NGS with parents of children with USH who were diagnosed later in life with the onset of vision problems. We hypothesize that there are implications, such as emotional issues or changes in decision-making, of early diagnosis of USH. The goals of this study are two-fold. The first goal is to assess the parental psychosocial implications of earlier diagnosis via genetic testing (before onset of RP) of Usher syndrome compared to later clinical diagnosis at onset of RP. The second goal of this study is to translate our results into genetic counseling (and/or other healthcare provider) conversations with parents pursuing use of hearing loss NGS panels or receiving a diagnosis of USH for their child.

## MATERIALS AND METHODS

The purpose of this study is to assess the psychosocial impact of early diagnosis of USH via NGS panels compared to the pre-NGS era where diagnosis was made later in life based on the onset of RP. This was done through a series of semi-structured interviews. The script used for interviews was developed using a variety of sources, including anecdotal evidence gathered from healthcare providers with experience with individuals with USH, as well as through a literature review exploring previous studies involving individuals affected by USH and implications of early diagnosis of other disorders. The final interview instrument included questions addressing topics such as feelings surrounding the diagnosis of both USH and future vision loss, impact of diagnosis on major life decisions and every-day life, and perception of appropriate timing of diagnosis. The final interview instrument was reviewed and approved by all authors. The full interview instrument is available by request.

84      *Caitlin Wright, Aimee Brown, Christina Hurst et al.*

Eligibility criteria for this study included being a parent of a child diagnosed with USH. All eligible participants were at least 18 years of age, English speaking, and were of any sex, race, and ethnicity. Subjects were recruited through UAB Ophthalmology and the Usher Syndrome Coalition (USC), a support and advocacy group for individuals affected by USH. Participation was emphasized as voluntary and confidential.

Telephone interviews were conducted by one author (CW) and lasted 30-90 minutes. Ages at diagnosis and method of diagnosis were collected to sort participants into two groups: parents of children diagnosed via genetic testing (and at earlier ages) and parents of children diagnosed via ophthalmologic findings (and at later ages). All interviews were recorded and labeled by participant number. Diagnosis via genetic testing generally correlated with younger ages and diagnosis via ophthalmologic findings generally correlated with older ages of diagnosis. No identifying information was collected during interviews and responses are not connected to any identifying information retrieved from medical records.

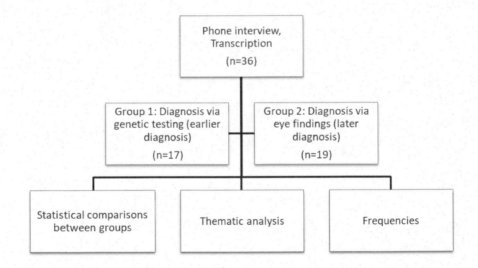

Figure 1. Workflow and analysis pipeline for participant responses.

Interviews for both study groups were transcribed using a RedCap online database. Thematic analysis was conducted by two authors to identify any common themes among participant responses. Any discrepancies in theme between the two authors were discussed and mutually agreed upon. Any participant responses that deviated from wording used in the database for transcription were recorded in the category that was the best fit. Themes that were identified were compared between the two groups using frequency and descriptive statistical analysis (See Figure 1 for workflow). Qualitative data that was collected during the interview was sorted into categories (i.e., positive vs. negative emotions regarding diagnosis) and analyzed using descriptive statistics and Chi-square analysis.

The study received approval from the Institutional Review Board of the University of Alabama at Birmingham.

# RESULTS

## Sample Characteristics

A total of 36 interviews, representing 35 different patients (both parents of 1 child were interviewed) were completed via phone and transcribed. All interview transcripts were separated into two groups based on method/age of diagnosis of the participants' children: genetic testing (n = 17), when individuals were typically diagnosed at younger ages (average age = 2.3 years) or ophthalmologic findings (n = 19), when individuals were typically diagnosed at older ages (average age = 15.9 years). Responses were grouped into categories, allowing for statistical analysis between the two groups. Chi square analysis, followed by Fisher's Exact Test was used to compare responses between the two groups. For questions related to preferred age of diagnosis, numerical values were grouped into the following categories: infancy (0-2 years), early-mid childhood (3-11 years), adolescence (12-18 years), and adulthood (19 years and above). More detailed demographic information for participants can be found in Table 2.

**Table 2. Demographic information separated by study group**

| Characteristic | Genetic Testing n (%) | Ophthalmologic Findings n (%) |
|---|---|---|
| Age | | |
| 25-30 | 3 (17.6) | -- |
| 31-35 | 2 (11.8) | -- |
| 36-40 | 9 (53) | 1 (5.3) |
| 41-45 | 3 (17.6) | 1 (5.3) |
| >45 | -- | 17 (89.5) |
| Race | | |
| African American | -- | -- |
| Asian | 1 (5.9) | -- |
| Caucasian | 15 (88) | 19 (100) |
| Hispanic | -- | -- |
| Other | 1 (5.9) | -- |
| Marital status | | |
| Single | 2 (11.8) | 1 (5.3) |
| Married | 15 (88) | 16 (84.2) |
| Divorced | -- | 2 (10.5) |
| Domestic partnership | -- | -- |
| Education level | | |
| Some high school | -- | -- |
| High school graduate | -- | 1 (5.3) |
| Some college | 2 (11.6) | -- |
| Trade/technical training | -- | -- |
| College graduate | 5 (29.4) | 8 (42) |
| Some postgraduate work | -- | 1 (5.3) |
| Postgraduate degree | 10 (59) | 9 (47.4) |
| Employment status | | |
| Employed | 14 (82.4) | 11 (57.9) |
| Unemployed | 3 (17.6) | 8 (42) |
| Total household income | | |
| Under $40,000* | 2 (11.8) | 3 (15.8) |
| $40,000 and above | 15 (88.2) | 16 (84.2) |
| Religious preference | | |
| Christian | 14 (82.4) | 11 (57.9) |
| Greek or Russian Orthodox | -- | -- |
| Jewish | -- | 1 (5.3) |
| Mormon | -- | -- |
| Muslim | -- | -- |
| Seventh Day Adventist | -- | -- |
| Other | 3 (17.6) | 7 (36.8) |

*Based on average household income in Alabama, USA (retrieved from: www.deptofnumbers.com).

## Emotional Impact of Diagnosis

Participants were asked about the types of emotions that they experienced surrounding the future onset or current vision loss experienced by their child. For analysis of qualitative data on emotions, participants' described emotions were grouped into three categories: positive, negative, and neutral. The majority (61%) of participants had negative emotions when they learned of the vision loss that is associated with USH. Examples of negative emotions included sadness, devastation, fear, and worry. A few participants (11%) reported experiencing positive emotions when they learned about the vision loss associated with USH. Examples of positive emotions included relief, hopefulness, strength, and empowerment. Over one quarter (28%) of participants experienced mixed (both positive and negative) or neutral emotions regarding vision loss that is associated with USH. There was no significant difference between study groups in types of emotions experienced with learning about the vision loss associated with USH $c^2$ (2, $N = 36$) = 0.3.2, $p = 0.2$. However, the number of participants in each group was too small to detect slight significant differences between the two groups.

Participants were also asked about emotions they experienced at the time of their child's diagnosis. For analysis, emotions were grouped into three categories: positive, negative, and neutral. The majority of all (81%) of participants experienced negative emotions when their child was diagnosed with USH. Around 14% of all participants experienced positive emotions when their child was diagnosed with USH, and 5% of all participants experienced neutral emotions. There was no significant difference between study groups in types of emotions experienced with diagnosis of USH $c^2$ (2, $N = 36$) = 0.12, $p = 0.94$. A summary of common themes among participant responses can be found in Table 3.

# Table 3. Common themes

| Common Theme | Example participant quote | % |
|---|---|---|
| Reassurance by healthcare providers regarding hearing loss was perceived negatively as providing false hope | "We were told at the hospital that it was probably just amniotic fluid in his ears, and it didn't seem like too much of a big deal. We weren't super concerned at that point." | 25% |
| Dissatisfaction with healthcare providers' delivery of diagnosis | "I think genetic counselors really need some grief counseling education, because they really can lay it on and not even think about how…at least it doesn't come off like they're even considering how that is being taken." | 33% |
| Diagnosis had a positive change on outlook on life | "I try to live in the moment as much as possible. I try to be grateful for as many things as you can: the ordinary every-day things that are going well." | 33% |
| Diagnosis is devastating, but still hopeful for a treatment or cure in the future | "They (physicians) told him that there are so many awesome treatments that are in trials at the moment. By the time he is an adult, he probably won't even have to worry." | 39% |
| Utilization of other parents of children with USH or affected individual for guidance and information | "There is a family that is close to me. They just started a foundation for vision and so they're really neat people. Their daughter is 6 and I know they are getting ready to talk to her so I kind of lean on them a lot because their daughter is a little older. I just kind of follow their lead with some things…" | 40% |
| Taking advantage of opportunities to make memories before vision loss is significant | "For our family, it's almost been a hidden blessing. We've travelled around the world. We've done thing with our kids that a lot of kids don't get to do and I've met so many amazing people." | 42% |
| Early diagnosis allows for more time for preparation and planning | | 42% |
| Early diagnosis allows more time to worry | "I think if we had known even earlier, I might have been worried all the time. Like when is it happening?" | 11% |
| Diagnosis inspired parents to help others affected by USH | "I just really believe strongly in folks who are deafblind. It's been a part of my life for a very long time. When you asked about careers and that sort of thing, much of what I do is volunteer. I had to do what I felt was right, and that was to be there for him and be involved in his community and be accepted by it." | 25% |
| Comparison of diagnosis to a life-threatening illness | "This is not going to kill her…then you feel grateful when you think of all of the awful conditions you could pass on to your children." | 28% |
| Diagnosis is a second wave of grief | "I got a little bit of it in the beginning, and then I got the big wallop a couple of years in. It's definitely like another kick in the gut, but I also think I had some coping skills at that time." | 31% |
| Diagnosis is a relief | "Before he got diagnosed, I was wracking my brain to figure out what I had done to cause the hearing loss. You definitely do a lot of that. That is one benefit of getting the genetic test results; that I was able to release myself from the guilt over the fact that maybe I had a couple of glasses of wine while I was pregnant." | 17% |

## Impact of Method/Timing of Diagnosis on Decision to Undergo Cochlear Implant or Early Intervention Services

All (100%) of the participants' children had hearing loss, which is a major characteristic of USH. The average age among both groups for diagnosis of hearing loss was 9.7 months of age (range: birth-5 years). Less than half (47%) of participants reported that their child had or planned to get a cochlear implant (CI). Of the participants who reported that their child had or planned to get a CI, just over half (53%) were in the group whose child was diagnosed via ophthalmologic findings. There was no significant difference between groups in the choice to pursue CI, $c^2$ (1, $N = 36) = 0.0003$, $p = 0.98$). Of the participants whose children have or plan to get a CI, the majority (83%) reported that the diagnosis of USH did not affect their decision to pursue CI. There was no significant difference between the two groups regarding the diagnosis of USH affecting plans to pursue CI, $c^2(1, N = 17) = 0.18, p = 0.67$.

Most of the participants (86%) reported pursuing early intervention (EI) services, such as occupational therapy, physical therapy, and speech therapy for their children. Of the participants who pursued EI services, over half (58%) belonged to the group of parents whose children were diagnosed via ophthalmologic findings. There was no statistically significant difference between the two groups regarding the decision to pursue EI services, $c^2(1, N = 36) = 2.5, p = .11$.

When asked if the USH diagnosis affected the decision to pursue EI services, the majority (77%) said that it did not affect their decision to pursue early intervention services, but two participants, who did not pursue EI, did not respond to this question. Of the participants who reported that the diagnosis did not affect their decision, over half (58%) were in the group whose child was diagnosed via ophthalmologic findings. Of the participants who reported that the diagnosis did affect their decision to pursue EI services (24%), over half (63%) were in the group whose child was diagnosed via genetic testing. There was no significant difference between the two groups' responses to whether USH diagnosis affected the decision to pursue EI, $c^2(1, N = 34) = 1, p = 0.32$.

## Impact of Early Diagnosis on Life Decisions

Participants were asked a series of questions about whether their child's diagnosis affected decisions they've made regarding family life. Topics included having additional children, career, place of residence, and schools for their child to attend. The majority (76%) of all participants reported that their child's diagnosis did not affect their decision to have future children. There was no statistically significant difference between the two groups' responses regarding the diagnosis affecting the decision to have future children $c^2$ (1, $N = 36$) = 1.8, $p = 0.18$. Half (50%) of participants reported that their child's diagnosis has affected decisions regarding their career, and half (50%) of participants reported that the diagnosis did not affect career decisions. There was no statistically significant difference between the two groups' responses regarding impact of diagnosis on career decisions $c^2$ (1, $N = 36$) = 0.11, $p = 0.74$. Half (50%) of participants reported that their child's diagnosis affected decisions about their place of residence, and half (50%) reported that the diagnosis did not affect decisions about their place of residence. There was no statistically significant difference between the two groups in response regarding whether the diagnosis affected decisions about place of residence $c^2$ (1, $N = 36$) = 1, $p = 0.32$. Over half (61%) of all participants reported that their child's diagnosis affected choices about schools for their child to attend. Around 39% of participants reported that the diagnosis did not impact decisions about their child's school. There was no statistically significant difference between groups' responses regarding the impact of diagnosis on decisions about schools to attend $c^2$ (1, $N = 36$) = 1.2, $p = 0.27$.

## Impact of Method of Diagnosis on Parental Perception of Preferred Timing of Diagnosis

Participants were asked if they felt their child was diagnosed with USH too late, too early, or at the right time. Responses were compared between the group of parents whose children were diagnosed via genetic testing and the group of parents whose children were diagnosed via ophthalmologic findings (See Figure 2). Many participants (83%) felt that their child was diagnosed at the right time. However, a few participants (8%) felt that their child was diagnosed too late, and a few (8%) participants felt that their child was diagnosed too early. There was a statistically significant difference in response between the two groups $c^2$ (2, $N$ = 36) = 6.0, $p$ = 0.049. All the participants who felt that their child was diagnosed too late (17% of the group) were in the group whose children were diagnosed via ophthalmologic findings, and therefore were diagnosed later in life. All the participants who felt that their child was diagnosed too early (18% of the group) were in the group whose children were diagnosed via genetic testing, and therefore were diagnosed earlier. Participants in the group whose children were diagnosed via ophthalmologic findings were more likely to answer that their child was diagnosed "too late" than participants in the group whose children were diagnosed via genetic testing. In contrast, individuals in the group whose children were diagnosed via genetic testing were more likely to answer "too early" than participants in the group whose children were diagnosed via ophthalmologic findings.

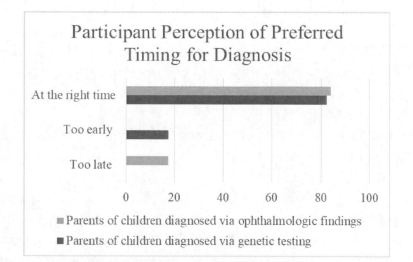

Figure 2. Parental perception of preferred diagnosis. This figure illustrates the percentage of participants from each group who felt that their children were diagnosed too early, too late, or at the right time.

## Impact of Method/Timing of Diagnosis on Frequency of Thoughts about Future Vision Loss

When participants were asked how frequently they think about the future vision loss that their child will experience, most (69%) reported that they think about it every day. Of the participants who answered, "every day," over half (60%) were in the group whose children were diagnosed via ophthalmologic findings. There was no statistically significant difference between the two groups in frequency of thought about future vision loss $c^2 (3, N = 36) = 5.7, p = 0.13$.

## Impact of Earlier Diagnosis on Recruitment for Participation in Clinical Trials

The majority (89%) of participants reported that their child has not been approached to participate in a clinical trial specifically for USH. Of the participants who reported their child had been approached to participate in a clinical trial (11% of all participants), 75% were in the group whose children were diagnosed via ophthalmologic findings. There was no statistically significant difference between the two groups in being approached to participate in clinical trials $c^2 (1, N = 36) = 0.89, p = 0.34$.

## Impact of Method/Timing of Diagnosis on Timing of Disclosure to Children about Vision Loss

Participants were asked at what age they planned to have or had the discussion about the future vision loss that occurs with USH (due to RP). Participants in the group whose children were diagnosed via genetic testing were more likely to respond with earlier ages, such as infancy, early, and mid-childhood, while parents in the group whose children were diagnosed via ophthalmologic testing were more likely to respond with later ages, such as adolescence or adulthood $c^2 (5, N = 36) = 23.1, p = .0003$.

## Opinions on the Most Appropriate Age for Diagnosis

Participants were asked, in their opinion, what is the most appropriate age for a child to receive the diagnosis of USH. Around 31% of all participants responded that it is most appropriate to receive the diagnosis of USH during adolescence, while 44% of participants felt that receiving the diagnosis as soon as possible was the most appropriate time. Around 12% of participants reported that infancy was the most appropriate age to receive the diagnosis of USH. The other 14% of participants reported that they did not know what the most appropriate age is for diagnosis of USH. Parents in the group whose children were diagnosed at later ages via ophthalmologic findings were more likely to respond that older ages are most appropriate for diagnosis $c^2 (3, N = 36) = 17.3, p = 0.0006$.

## Impact of Method/Timing of Diagnosis on Parental Changes in Overall Outlook on Life

The majority (61%) of participants reported that their child's diagnosis has led to a positive change in overall outlook on life. Around 19.4% of participants reported a negative change and 19.4% of participants reported no change in their overall outlook on life. Of the participants who reported a positive change in outlook, over half (55%) were in the group whose child was diagnosed via genetic testing. There was no significant difference between groups in change in overall outlook $c^2 (2, N = 36) = p = 0.15$.

## Conclusion

The aim of this study was to evaluate whether there is a greater psychosocial impact of early diagnosis of Usher syndrome (USH) via Next-Generation Sequencing (NGS) panels compared to the pre-NGS era where diagnosis was made later in adolescence via ophthalmologic findings. We found that earlier diagnosis does not cause an increase in psychosocial issues for parents. This was surprising, as we expected that earlier diagnosis would be associated with increased psychosocial issues, such as anxiety and distress. There was no significant difference in types of emotions experienced at the time of diagnosis between parents whose child was diagnosed at a young age via NGS panel and parents whose child was diagnosed at a later age via ophthalmologic findings. This has significant implications, as the concerns about the negative effects with earlier diagnosis are apparently unfounded, at least in this population. Therefore, earlier diagnosis can be viewed as primarily beneficial.

A common response of parents who preferred earlier diagnosis was that earlier diagnosis of USH allows for more time for preparation for planning in the sense of education, finances, and support services. One participant in particular mentioned that the diagnosis is still a negative experience, but knowing earlier is beneficial, because it can allow for planning: *"Back to your basic question, which is: Is it better to tell parents when their child is a newborn and they fail the test and then you get a blood test and it says that you have Usher? Ab-so-freaking-lutely. The reason why is you're going to grieve anyway and it's going to change your life anyway, but when you know what you're dealing with, you can plan for it."* This result concurs with research by Luterman and Kurtzer-White (1999) that explored the impact of newborn hearing screening programs and demonstrated parents' desire for earlier diagnosis of hearing loss in order to plan and access services earlier. Multiple participants also mentioned that earlier diagnosis allows them to be proactive in making memories and unique experiences with their children while the child still has their vision. Examples included anything from waking a child up from a nap to see a rainbow to taking their child on trips to meet foreign dignitaries. At least half of all participants reported that they changed life plans regarding career, place of residence, and schools for their child based on the diagnosis of USH. This suggests that USH has a major impact on both the affected individual's and their family's lives. Although parents in the group whose children were diagnosed early were not more likely to change life plans based on diagnosis, earlier diagnosis might in fact allow for more planning and time to make decisions regarding career, place of residence, and schools to attend.

Thus, one of the obvious benefits of earlier diagnosis of USH is the ability to prepare and plan for many aspects of a child's life before the onset of vision loss.

As might be predicted, diagnosis of USH at any time during a child's life brings up negative emotions such as fear and devastation. However, a few parents specifically mentioned experiencing positive emotions when receiving the diagnosis for their child at an earlier age. Specifically, one participant mentions a relief of guilt after being told her child's hearing loss was due to a genetic cause:

> "Before he got diagnosed, I was wracking my brain to figure out what I had done to cause the hearing loss. You definitely do a lot of that. That is one benefit of getting the genetic test results; that I was able to release myself from the guilt over the fact that maybe I had a couple of glasses of wine while I was pregnant."

Earlier diagnosis allows for the relief of maternal guilt (believing they caused their child's hearing loss) much sooner than if a child is diagnosed in late childhood with USH, which could affect the parent's overall coping with the hearing loss as well as with the vision loss and USH diagnosis itself.

As one might expect, the majority of all participants thought about the future vision loss their child would experience daily, regardless of method/age of diagnosis. This suggests that the diagnosis of USH does have a significant impact on the lives of parents. However, earlier diagnosis seems to give parents more time to consider the future vision loss and cope with it before having to disclose this information to their affected child. With earlier diagnosis of USH, parents also have the ability to take time to grieve and process the diagnosis to feel comfortable with the diagnosis before they are tasked with supporting their child through his or her grief for the vision loss. In fact, many parents reported that after coping with the diagnosis, they found they had a positive change on their outlook on life. They learned to reprioritize their lives and to appreciate more. They also learned valuable ways to cope, such as forming relationships with other parents of children with USH or individuals with vision loss. Earlier diagnosis even inspired many parents to help others by beginning foundations to help promote research or promote special events to make memories. There were more benefits than detriments of earlier diagnosis via genetic testing mentioned by participants.

Both groups of parents pursued CI and early intervention services regardless of the timing/method of their child's diagnosis of USH. However, many parents mentioned that they used the diagnosis of USH to get an additional CI or to pursue more services such as additional visual screening. Another potential benefit of earlier diagnosis of USH is that it can be helpful in getting services like bilateral CI and earlier intervention services into place more quickly. If parents are able to find out about the future vision loss (due to RP) sooner, they might be able to get their child into services to help with vision issues. If affected individuals are able to learn strategies to communicate and travel independently even after onset of RP, this may lead to more independence throughout their lives, which would complement findings by Zannon de Andrade Figueiredo, et al. (2013) that adults with USH wished that they had received services for visual impairment before the onset of vision loss.

Interestingly, there was a significant difference between groups regarding perception of whether their child was diagnosed too late, too early, or at the right time. This significant difference is not surprising, given that people in both groups had negative experiences receiving the diagnosis. However, the clear majority of participants felt that their child was diagnosed at the right time. This finding suggests that earlier diagnosis via genetic testing does not have a wholly negative impact on parents. In fact, it suggests that parents are able to cope with earlier diagnosis, given that 82% of participants in the group whose children were diagnosed via genetic testing responded that they felt the diagnosis was at the right time.

The significant difference between proposed ages to discuss the vision loss with the child mirrors the experiences of the participants in each group. Participants whose children were diagnosed later via ophthalmologic findings felt more comfortable talking about the vision loss with their child at later ages, because it was affecting their child at the time of the discussion. For the most part, participants whose children were diagnosed earlier via genetic testing felt more comfortable talking to their child about the impending vision loss at younger ages, so that they maintained an open conversation with the child as the child grew. This result suggests that earlier diagnosis may allow parents to feel more comfortable discussing vision loss with their child before its onset.

Another potential benefit of early diagnosis, via genetic testing, of USH is the ability for more individuals with USH to be identified for research purposes. Many participants seemed interested in research but were unable to find a research study for which their child is eligible. If more individuals with USH are identified, especially at younger ages, this provides an opportunity to study and better understand the natural history of the various types of USH. Earlier diagnosis via genetic testing may also provide a means to understand more about the genetic basis of USH and how specific genotypes may produce certain phenotypes through research. Although earlier diagnosis may allow for observational research, it could also help with producing studies that aim to cure RP and USH. One exciting innovation in genetic research is the use of CRISPR-Cas9 to perform gene modification. This genome-editing tool could prove to be valuable in the treatment of genetic disease, including USH. If this tool ever becomes available for use in the treatment of human disease, earlier diagnosis of USH would allow parents of these individuals to potentially utilize this tool before the onset and/or progression of significant vision loss.

## STRENGTHS AND LIMITATIONS

This is the first study, to our knowledge, to assess the impact of Usher syndrome diagnosis on parents of affected children, as well as to assess the emotional implications of the timing of this diagnosis (now available with next-generation sequencing panels). The results of this study provide evidence that earlier diagnosis, through genetic testing, does not have a significantly more negative impact on parents than later diagnosis of USH, previously delayed until onset of vision loss. This is a small study, with 36 participants. However, it is a large sample for a qualitative research study and provides rich data about parental perspectives and feelings. The majority of participants were Caucasian, well-educated and of a higher socio-economic status. This may make results less generalizable to the general population.

## PRACTICE IMPLICATIONS

One prominent theme from this study was the dissatisfaction with how healthcare providers disclosed the diagnosis of USH to parents, regardless of the setting of the provider. It is important that all healthcare providers use this information to adjust how they disclose results to families. It may be necessary to provide more counseling before genetic testing to forewarn parents of the possibility of a positive result for a disorder that has future implications to ensure that parents are truly providing informed consent. When delivering the diagnosis of USH syndrome, healthcare providers should anticipate possible emotions that parents might feel and should be prepared to provide support to parents. Additional training in how to handle grief or how to deliver bad news might be necessary for healthcare providers to deliver the best possible support to parents of children diagnosed with USH syndrome.

Another theme that came up regarding the experience surrounding the diagnosis was the lack of information and appropriate resources for parents to utilize as they processed the information. Healthcare providers should be aware of the extreme emotional impact on parents that diagnoses such as hearing loss and USH have on parents. All healthcare providers should be prepared to provide information for support groups, such as the Usher Syndrome Coalition, to parents of a newly diagnosed child. Even healthcare providers who follow individuals with USH for long periods of time, such as audiologists and ophthalmologists should be aware of the support groups and resources that are available to families with individuals affected with USH. Although the natural history of USH is still not well understood, it would be helpful to provide informational resources to parents on the specific type of USH that the child has.

## FUTURE RESEARCH

Additional studies into the psychosocial implications of earlier diagnosis of USH syndrome with a larger sample would be useful in providing even more insight into additional benefits of early diagnosis. Validated scales to measure stress would be useful to assess whether parents of children with USH diagnosed at earlier ages experience more stress than parents of children diagnosed at older ages. Additionally, there are very few studies on the experiences of individuals who are affected with Usher syndrome. It would be interesting and useful to determine their feelings regarding earlier diagnosis of Usher syndrome via genetic testing.

A few participants from this study mentioned that they had not yet disclosed their child's diagnosis to their pre-teen children. Future studies assessing the impact on individuals with USH of early versus late disclosure of the diagnosis by their parents would provide useful information regarding whether earlier diagnosis has major psychosocial implications for these individuals.

## CONCLUSION

Early diagnosis of Usher syndrome via genetic testing does not have psychosocial implications that are significantly different from the later diagnosis via traditional ophthalmologic findings. To the contrary, the results of this study suggest that there are multiple benefits to early diagnosis. Diagnosis of Usher syndrome early in a child's life allows parents to process emotions regarding the diagnosis and prepare the child for independence in the future. Earlier diagnosis of Usher syndrome not only allows researchers to identify more potential participants for Usher syndrome specific studies, but also allows for parents to join support groups and be vigilant for research and clinical trials that could be helpful to their child in the future. The results of this study suggest that the benefits of earlier diagnosis outweigh the negative emotions that parents experience immediately after diagnosis.

## REFERENCES

Alford RL, Arnos KS, Fox M, Lin JW, Palmer CG, Pandya A, Rehm HL, Robin NH, Scott DA, Yoshinaga-Itano C. 2014. "ACMG Working Group on Update of Genetics Evaluation Guidelines for the Etiologic Diagnosis of Congenital Hearing Loss; Professional Practice and Guidelines Committee. American College of Medical Genetics and Genomics guideline for the clinical evaluation and etiologic diagnosis of hearing loss." *Genetics in Medicine* 16(4):347-55. doi: 10.1038/gim.2014.2.

Brabyn JA, Schneck ME, Haegerstrom-Portnoy G, Lott LA. 2007. "Dual sensory loss: overview of problems, visual assessment, and rehabilitation." *Trends in Amplif*ication 11(4):219-26.

Cohen, M., Bitner-Glindzicz, M., and Luxon, L. 2007. "The changing face of Usher syndrome: Clinical implications." *International Journal of Audiology 46*(2), 82-93. doi:10.1080/14992020600975279.

Damen, Godelieve W. J. A., Krabbe, Paul F. M., Kilsby, M., Mylanus, Emmanuel A. M. 2005. "The Usher lifestyle survey: Maintaining Independence: a Multi-centre Study." *International Journal of Rehabilitation Research* 28(4) 309-320.

Dammeyer J. 2014. "Deafblindness: a review of the literature." *Scandinavian Journal of Public Health* Nov;42(7):554-62. doi: 10. 1177/1403494814544399.

Decarlo DK, McGwin G Jr, Bixler ML, Wallander J, Owsley C. 2012. "Impact of pediatric vision impairment on daily life: results of focus groups." *Optometry and Vision Science* Sep;89(9):1409-1416.

Ellis, L., and Hodges, L. 2013. *The experiences of diagnosis for people with Usher syndrome* [Scholarly project]. In *University of Birmingham*. Retrieved from http://www.birmingham.ac.uk/schools/ education/research/victar/research/usher-syndrome.aspx.

Fahim AT, Daiger SP, Weleber RG. Retinitis Pigmentosa Overview. 2000 Aug 4 [Updated 2013 Mar 21]. In: Pagon RA, Adam MP, Ardinger HH, et al., editors. GeneReviews® [Internet]. Seattle (WA): University of Washington, Seattle; 1993-2016. Available from: https:// www.ncbi.nlm.nih.gov/books/NBK1417/.

Idan, N., Brownstein, Z., Shivatzki, S., and Avraham, K. B. 2013. "Advances in genetic diagnostics for hereditary hearing loss." *Journal of Basic Clinical Physiology and Pharmacology* 24(3), 165-170. doi:10.1515/jbcpp-2013-0063.

Lentz J, Keats BJB. Usher Syndrome Type I. 1999 Dec 10 [Updated 2016 May 19]. In: Pagon RA, Adam MP, Ardinger HH, et al., editors. GeneReviews® [Internet]. Seattle (WA): University of Washington, Seattle; 1993-2016. Available from: https://www.ncbi.nlm.nih.gov/ books/NBK1265/.

Lentz J, Keats B. Usher Syndrome Type II. 1999 Dec 10 [Updated 2016 Jul 21]. In: Pagon RA, Adam MP, Ardinger HH, et al., editors. GeneReviews® [Internet]. Seattle (WA): University of Washington, Seattle; 1993-2016. Available from: https://www.ncbi.nlm.nih.gov/ books/NBK1341/.

Luterman, D. 1999. "Counseling Families with A Hearing-Impaired Child." *Otolaryngologic Clinics of North America* 32(6), 1037-1050. doi:10.1016/s0030-6665(05)70193-6.

Luterman, D., and Kurtzer-White, E. 1999. "Identifying Hearing Loss." *American Journal of Audiology* 8(1), 13. doi:10.1044/1059-0889 (1999/006).

Mellon, N. K., Niparko, J. K., Rathmann, C., Mathur, G., Humphries, T., Napoli, D. J., Lantos, J. D. 2015. "Should All Deaf Children Learn Sign Language?" *Pediatrics 136*(1), 170-176. doi:10.1542/peds.2014-1632.

Mitchell, R., and Karchmer, M. 2004. "Chasing the mythical ten percent: parental hearing status of deaf and hard of hearing students in the United States." *Sign Language Studies* 4(2): 138-163.

Sloan-Heggen, C. M., and Smith, R. J. 2016. "Navigating genetic diagnostics in patients with hearing loss." *Current Opinion in Pediatrics* 1. doi:10.1097/mop.0000000000000410.

Szarkowski, A., and Brice, P. J. 2016. "Hearing Parents' Appraisals of Parenting a Deaf or Hard-of-Hearing Child: Application of a Positive Psychology Framework." *Journal of Deaf Studies and Deaf Education DEAFED 21*(3), 249-258. doi:10.1093/deafed/enw007.

Vona B, Müller T, Nanda I, Neuner C, Hofrichter MA, Schröder J, Bartsch O, Läßig A, Keilmann A, Schraven S, Kraus F, Shehata-Dieler W, Haaf T. 2014. "Targeted Next-Generation Sequencing of Deafness Genes in Hearing-Impaired Individuals Uncovers Informative Mutations." *Genetics in Medicine* vol. 16. doi:10.1038/gim.2014.65.

Zannon de Andrade Figueiredo, Marilia, Chiari, Brasilia Maria, Garcia de Goulart, Barbara Niegia. 2013. "Communication in deafblind adults with Usher syndrome: retrospective observational study." *Communication Disorders, Audiology, and Swallowing* 25(4), 319-324.

In: A Comprehensive Guide …  ISBN: 978-1-53616-975-1
Editor: Benjamin A. Kepert  © 2020 Nova Science Publishers, Inc.

*Chapter 3*

# GENETIC COUNSELING (GC) IN THE ARAB SOCIETY OF ISRAEL

*Abdelnaser Asad Zalan[1,]\*, PhD*
*and Rajech Abdullah Sharkia[1,2], PhD*
[1]Department of Human Biology,
The Triangle Regional Research and Development Centre,
Kfar-Qari', Israel
[2]Beit-Berl Academic College, Beit-Berl, Israel

## ABSTRACT

Genetics is considered to be a scientific study of the mechanism of inheritance and causes of variation in all of the living organisms related by descent. Although man has always been aware that different individuals vary between themselves but children do resemble their parents in some way. The scientific basis for these observations was given thrust during the past 150 years. The clinical application of this knowledge has become widespread with most progress made in the past few decades. Genetic diseases were known long ago, but they were poorly understood. A more specific measure of the impact of genetic

\* Corresponding Author's E-mail: dr.zalan@hotmail.com.

disorders is their role in mortality and morbidity. In an attempt to cope with these genetic diseases, the process of genetic counseling (GC) was newly introduced. Thus, GC could be defined as the process of advising individuals and families affected by or at risk of genetic disorders to help them understand and adapt to the medical, psychological and familial implications of genetic contributions to disease.

The advent of prenatal diagnosis and the ability to detect an inborn error of metabolism, chromosome disorder, or congenital malformation in utero, early enough to allow termination of pregnancy, has added a new dimension to genetic counseling.

Almost, all Arab countries are characterized by high rates of consanguineous marriages with a common founder effect and do have common and rare gene-mutations that give rise to genetic disorders. As the Arab society of Israel has unique socio-demographic and cultural characteristics with some common and rare genetic disorders and high rates of consanguineous marriages, that differ from the rest of the population, therefore, genetic counseling for this population should be carefully designed.

We report on a comprehensive analysis of the local studies performed so far on genetic counseling particularly those pertaining to the Israeli Arab society. This review is an attempt to locate certain procedures that should be implemented while extending the genetic counseling services to all targeted population. Furthermore, an overview of the various socio-demographic, economic, cultural and religious aspects will be taken into account.

**Keywords**: Arab society in Israel, genetic counseling, genetic disorders, pre-natal genetic testing, risk factors of genetic diseases

## ABBREVIATIONS

| | |
|---|---|
| AC | Amniocentesis |
| Allo-HSCT | Allogeneic hematopoietic stem cell transplantation |
| ASHG | American Society of Human Genetics |
| CBS | Central Bureau of Statistics |
| CF | Cystic fibrosis |
| CFTR gene | Cystic fibrosis transmembrane-conductance regulator gene |

| | |
|---|---|
| CM-AVM syndrome | Capillary malformation-arteriovenous malformation syndrome |
| CVS | Chorionic villus sampling |
| DNA | Deoxyribonucleic acid |
| FMF | Familial Mediterranean Fever |
| GC | Genetic counseling |
| GCDTF | Genetic Counseling Definition Task Force |
| GD | Gaucher disease |
| HGMD | Human gene mutation database. |
| HMW | High-molecular-weight |
| LOVD | Leiden open variation database |
| MEFV | Mediterranean fever |
| MOH | Ministry of Health |
| NGS | Next-generation sequencing |
| NIPD | Non-invasive prenatal diagnosis |
| NSGC | National Society of Genetic Counselors |
| NTD | Neural tube defect |
| OECD | Organization for Economic Co-operation and Development |
| PGD | Preimplantation genetic diagnosis |
| PGT-SR | Preimplantation genetic testing-structural rearrangements |
| STRs | Short tandem repeats |
| SUD | Substance use disorder |
| UK | United Kingdom |
| UNESCO | United Nations Educational Scientific and Cultural Organization |
| US | Ultrasound |
| WHO | World Health Organization |

# 1. INTRODUCTION

## 1.1. General Description of Genetics and Genetic Services

Genetics is considered to be a scientific study of the mechanism of inheritance and causes of variation in all of the living organisms related by descent. Although man has always been aware that different individuals vary between themselves but children do resemble their parents in some way. The scientific basis for these observations was given a thrust during the past 150 years. The clinical application of this knowledge has become widespread with most progress made in the past few decades (Grabber and Vanarsdall, 2005). Human genetics has been used widely and experienced significant advances for prevention of illness and occasionally for treatment (Caulfield, 1998). Recent advances in genomics technology such as high throughput sequencing have contributed to a better diagnosis and understanding of the genetic basis of inherited disorders, and brought the prospect of prevention and tailor-made treatment closer to fruition (Hong and Oh, 2010). In the second half of the twentieth century, anthropological genetics has emerged as a new discipline to investigate the origin of human species. Anthropological geneticists started redrawing the ancient migratory history of human populations by using the genetic database of blood groups and other protein polymorphisms. South Korean physicians produced the genetic knowledge and discourse of the Korean origin (Hyun, 2019).

A report, from the world health organization (WHO) Genomics and World Health, states that the importance of developing genetic services worldwide, including public education and genetic counseling, cannot be overemphasized. The report also warns against introducing medical genetic services in the absence of public education and without the support of genetic counselors (WHO, 2002). A WHO scientific group, in 1993, endorsed the following principles of genetic counseling: (a) the autonomy of the individual or couple, (b) their right to full information, and (c) strict confidentiality of genetic test results (WHO Scientific Group on the Control of Hereditary Diseases, 1993).

Historically, clinical genetic services were developed in major hospitals in large urban centers, in both developed and developing countries, to provide diagnosis and counseling to patients with, or at risk for, genetic diseases and congenital anomalies. In developing countries, the major bottlenecks to increasing access to genetic services have been fundamentally the shortage of trained specialists, the lack of appropriate technology, the scarcity of funding, and the lack of links with the primary health care level (Ballantyne et al., 2006).

Naturally occurring microchimerism refers to the bidirectional interaction between the genetically distinct mother and fetus(es) resulting in the long-lasting presence of small numbers (well below 1%) of fetal cells in the mother and vice versa. The interesting immunological consequences of these coexistences have recently been elegantly reviewed (Eikmans et al., 2014). So far, more than 700,000 STRs (short tandem repeats) are characterized in human genome, for the most part they are considered as 'junk DNA,' however some of them are located in protein coding region and associated with genetic disorders (International Human Genome Sequencing Consortium, 2001; Mirkin, 2007; Willems et al., 2014). STRs—also referred to as microsatellites—consist of 2 to 6 nucleotide tandem repeats distributed throughout the genome. These loci are highly polymorphic and different alleles are determined by the varying number of nucleotide repeats (3 to 51 repeats). STR analysis is a standard and approved method applied in forensics and in hematopoietic chimerism quantification after allo-HSCT (Gettings et al., 2015). Nearly 100% informativity can be reached with a panel of 12–16 STR markers (Core STR Loci), including loci on sex chromosomes, that allow differentiation between individuals (Butler, 2007; Clark et al., 2015; Fan and Chu, 2007).

One of the recent studies found that, usually, mutations causing severe damage to proteins cause congenital cataracts, while milder variants increasing susceptibility to environmental insults are associated with age-related cataracts. The study concluded that new therapeutic approaches to age-related cataracts use chemical chaperones to solubilize the high-molecular-weight (HMW) protein aggregates, while attempts are being made to regenerate lenses using endogenous stem cells to treat congenital cataracts (Shiels and Hejtmancik, 2019).

## 1.2. General Description of Genetic Disorders

There are rare genetic diseases that are generally present in high frequency in certain communities worldwide as a result of some sort of isolation they suffer from. This isolation may be basically due to geographical, political, social, ethnic, economic or religious reasons. This leads to a limited variation in human relations leading to religious or consanguineous marriages. The relationship of consanguinity to inherited genetic disorders in the offspring was generally known since the old ages. However this relationship is deeply explored nowadays due to recent advances in molecular genetics (Shalev, 2019).

Genetic diseases are the outcome of chromosomal abnormalities. They occur mostly as a result of biological errors in the transmission of chromosomes from parents to offspring. The factors that influence the prevalence of genetic diseases in populations and their impact on the public's health are best understood by specific category of disease. The average birth prevalence of chromosomal abnormalities in live born infants is 1 in 200, and the most common example is Down syndrome. The main clinical manifestations of chromosomal abnormalities are congenital anomalies, mental retardation and infertility. Further, they are the main cause of miscarriages, as the overwhelming majority of chromosomally abnormal embryos are aborted spontaneously (an estimated 60% of spontaneous abortions involve a fetus affected by a chromosome abnormality). It is estimated that chromosomal abnormalities are responsible for about 10–20% of congenital anomalies and mental retardation in children (WHO, 1996).

A more specific measure of the impact of genetic disorders is their role in mortality and morbidity. Studies of children's hospitals which are referral in nature show a high percentage of genetic diseases among the causes of death. As for morbidity, it was found in Ireland that 26% of all institutional beds, 6% of all consultations with general practitioners, and 8% of those with specialists were for patients with genetically determined disease. A study in the United States of all admission diagnoses to a pediatric service in a university hospital showed that 7.1% of all admissions were for diseases of clearly defined genetic origin and 31.5% were possibly gene influenced (Sutton, 1998).

A recent study was conducted on Swedish population in order to assess whether parental substance use disorder (SUD) is associated with lower cognitive ability in offspring, and whether the association is independent of shared genetic factors. The study concluded that there appear to be shared genetic factors between parental SUD and offspring cognitive function, suggesting that cognitive deficits may constitute a genetically transmitted risk factor in SUD (Khemiri et al., 2019).

Another recent study was conducted in China which was intended to report the normal live birth and birth defect rates pre- and post-preimplantation genetic testing for chromosomal structural rearrangements (PGT-SR) in reciprocal translocation carriers who have experienced two or more unfavorable pregnancy histories. The study concluded that after the PGT-SR treatment, the reciprocal translocation carriers who had previously experienced two or more unfavorable pregnancy outcomes had a low risk of miscarriages and birth defects. The rate of normal live births per pregnancy was higher after PGT-SR treatment (Huang et al., 2019).

## 1.3. Prenatal Diagnosis

The advent of prenatal diagnosis and the ability to detect an inborn error of metabolism, chromosome disorder, or congenital malformation in utero, early enough to allow termination of pregnancy, has added a new dimension to genetic counseling. In place of confronting parents with the choice of planning further children in full knowledge of high risk of abnormality or limiting their families, it is now possible in a rapidly growing number of disorders to monitor at-risk pregnancies and ensure that children born will not be affected (Brock, 1974).

Genetic counseling fills a distinctive position in the complicated and varied arena of genomic medicine. Advances in genetic medicine create an even greater demand for expert health-care services. Genetic counselors help meet this need, serving in almost every major medical center and across the globe as an increasingly important resource for medical referral and quality patient care (Sutton, 1998; Phillips, 2001). It is important to cautiously utilize the scarce resources for prevention and care of genetic and congenital conditions (Ballantyne et al., 2006). The development of prenatal screening and prenatal diagnosis is reducing the prevalence of chromosomal abnormalities in countries where these services are available and accessible, as they are in developed countries (Chaabouni et al., 2001). It is well-known that disorders of genetic etiology exist in 2%-3% of live-born infants, therefore, identifying couples with increased susceptibility for offspring with anomalies or genetic disorders is increasingly effective as a result of molecular advances (Simpson et al., 2019). The non-invasive prenatal diagnosis (NIPD) for monogenic disorders has a high uptake by families in the United Kingdom (UK). One of the recent studies, used definitive NIPD for cystic fibrosis (CF). The study concluded that timely and accurate NIPD for definitive prenatal diagnosis of CF is possible in a public health service laboratory (Chandler et al., 2019).

## 1.4. History and Origin of Genetic Counseling (GC)

The term "genetic counseling" was first introduced by Sheldon Reed, a geneticist at the University of Minnesota, in 1947, who described the new field of genetic counseling as a type of social work (Reed, 1974). Sheldon expanded his views of genetic counseling in several issues of the Dight Institute Bulletin and also presented the concept at the First International Congress of Human Genetics, in Copenhagen, in 1956. It was his hope that genetic counseling would continue to be of help to individual families and that it would not become the tool of any governmental population program. He also thought that its future would be of the greatest interest, particularly in relation to ethical concepts (Reed 1979). Sheldon personally handled well over 4,000 cases of genetic counseling. When individuals or families came to him, he listened carefully and spoke simply to give them the information they would need to make their own decisions. He insisted that the presentation of this genetic information "must be compassionate, clear, relaxed, and without a sales pitch." He believed that the counselor also could help to alleviate some difficulties associated with genetic problems: quarreling between husband and wife as to the "blame" for an abnormality in their child and a sense of shame owing to the social stigmas that often accompany hereditary diseases. Maternal guilt is an emotional reaction that should be watched for. On the other hand, one usually can explain to parents the chances of another abnormal child, so that they can adjust well to the facts (Anderson, 2003).

Since this first description, many definitions of genetic counseling have been proposed. In 1974, genetic counseling was succinctly defined by Fraser as "a communication process that deals with the human problems associated with occurrence or risk of occurrence of a genetic disorder in a family" (Fraser, 1974). This was further expanded in 1975 by an ad hoc committee reporting to the American Society of Human Genetics (ASHG). This committee developed the following definition of genetic counseling that was broadly utilized for over 30 years: A communication process which deals with the human problems associated with occurrence, or the risk of occurrence, of a genetic disorder in a family. This process involves an attempt by one or more appropriately trained persons to help the individual or family to:

- Comprehend the medical facts, including the diagnosis, probable cause of the disorder, and the available management,
- Appreciate the way heredity contributes to the disorder, and the risk of recurrence in specified relatives,
- Understand the alternatives for dealing with the risk of recurrence,
- Choose the course of action which seems to them appropriate in view of their risk, their family goals and their ethical and religious standards, and to act in accordance with that decision and;

- Make the best possible adjustment to the disorder in an affected family member and/or to risk recurrence of that disorder.

Genetic counseling also implies a definition that is related to education, as an educational process that seeks to assist affected and/or at risk individuals to understand the nature of the genetic disorder, its transmission and the options open to them in management and family planning (Kelly, 1986). Additionally, genetic counseling is considered to be a basic and an indivisible part of the delivery of genetic services. Genetic counseling has been defined as: the process by which patients or relatives at risk of a disorder that may be hereditary are advised of the consequences of the disorder, the probability of developing or transmitting it and the ways this may be prevented, avoided or ameliorated (Harper, 2004).

While the ASHG definition captures many important aspects of genetic counseling it lacks specific reference to the counseling elements (Biesecker and Peters, 2001). The definition has been described as lengthy and complex – and more critically, does not reflect changes in medical care and genetics that have occurred during the past 30 years (Resta, 2006).

In 2006, the Genetic Counseling Definition Task Force (GCDTF) of the National Society of Genetic Counselors (NSGC) (Resta et al., 2006) developed a slightly more succinct and modern definition which is:

"Genetic Counseling is the process of helping people understand and adapt to the medical, psychological, and familial implications of genetic contributions to disease." This process integrates the following:

1. Interpretation of family and medical histories to assess the chance of disease occurrence or recurrence.
2. Education about inheritance, testing, management, prevention, resources and research.
3. Counseling to promote informed choices and adaptation to the risk or condition.

This definition emphasizes that genetic counseling is a psycho-educational process centered on the provision of genetic information (Kessler, 1997) with the goal being to facilitate a client's ability to use genetic information in a personally meaningful way in order to minimize psychological distress and increase perceived personal control (Biesecker and Peters, 2001).

## 2. GENETIC COUNSELING WORLDWIDE

### 2.1. General Description

The principles of informed consent in the context of genetic testing are upheld in several human rights acts. Obtaining informed consent to genetic testing and screening is essential to promote individual autonomy in medical decision-making. The process of gaining informed consent aims to ensure that the individual understands the relevant information (medical, social and/or legal) and that their decision to undergo any medical intervention is made voluntarily. Valid informed consent for genetic testing requires a bilateral process involving a dialogue of questions and answers between the individual considering testing and the person obtaining informed consent (often a health care professional). This dialogue requires the person obtaining informed consent to gauge the appropriate level of language and technical detail suitable for the individual's understanding (UNESCO, 1999; 2003).

Genetic counseling is a communication process, which aims to help individuals, couples and families understand and adapt to the medical, psychological, familial and reproductive implications of the genetic contribution to specific health conditions. This definition was developed by the Genetic Counseling Definition Task Force of the National Society of Genetic Counselors (NSGC). This process integrates the following: Interpretation of family and medical histories to assess the chance of disease occurrence or recurrence. Education about the natural history of the condition, inheritance pattern, testing, management, prevention, support resources and research. Counseling to promote informed choices in view of risk assessment, family goals, ethical and religious values. Support to encourage the best possible adjustment to the disorder in an affected family member and/or to the risk of recurrence of that disorder (Resta et al., 2006).

Genetic counseling for patients originating from isolated populations can be especially challenging. High rates of consanguineous marriage markedly increase the frequency of unexpected rare autosomal recessive disorders. Endogamous marriage involves at least 10% of the world population, especially common in the Middle East, North and Sub-Saharan Africa, and Central and South Asia (Becker et al., 2015).

One of the genetic disorders that is inherited in an autosomal dominant manner, is called capillary malformation-arteriovenous malformation (CM-AVM) syndrome, which is characterized by the presence of multiple small (1-2 cm in diameter) capillary malformations mostly localized on the face and limbs. This type of inheritance refers to a trait or disorder in which the phenotype can be expressed in individuals who have one copy of a pathogenic variant at a particular locus (heterozygotes); specifically refers to a gene on one of the 22 pairs of autosomes- non-sex chromosomes (Bayrak-Toydemir et al., 2019).

When we talk about genetic counseling (GC) there are certain terms and definitions that common people should be familiar with. Therefore, common people may have certain explanatory questions that should be answered in order to get the required help they need. Such questions could be about: genetic professionals, their work, their place of existence, their sub-specialization, other specialists who can assist them, etc. The genetic professionals are health care professionals with specialized degrees and experience in medical genetics and counseling, they include geneticists, genetic counselors and genetics nurses. They work as members of health care teams providing information and support to individuals or families who have genetic disorders or may be at risk for inherited conditions. These genetic professionals perform many tasks such as:

- Assess the risk of a genetic disorder by researching a family's history and evaluating medical records.
- Weigh the medical, social and ethical decisions surrounding genetic testing.
- Provide support and information to help a person make a decision about testing.
- Interpret the results of genetic tests and medical data.
- Provide counseling or refer individuals and families to support services.
- Serve as patient advocates.
- Explain possible treatments or preventive measures.
- Discuss reproductive options, (Abacan et al., 2018).

Due to large waves of immigration from areas with high rates of inbreeding to Europe, America and Oceania, consanguinity has become a global rather than a local phenomenon. Additional obstacles commonly encountered in isolated population are low awareness of the importance and benefits of prenatal genetic counseling, a possible tendency to obscure a severe medical condition to prevent stigmatization, a language barrier, and inferior documentation (Shalev, 2019).

In France, a study was conducted to analyze the indications and the results of invasive testing for fetal karyotyping for ultrasound abnormality in the third trimester of pregnancy, when first- and second-trimester screening tests were negative. The study concluded that, in low risk patients, fetal karyotyping in the third trimester may be justified when the diagnosis of fetal malformation is made in the third trimester of pregnancy. It was further noted that two or more anomalies increase the risk of fetal aneuploidy even with a negative-screening test in the first and second trimester of pregnancy (Drummond et al., 2003). A study was conducted in Germany, to assess the risk of major anomalies in the offspring (35,391 fetuses) of consanguineous couples (over a 20 year period) from 1993 to 2012. When the fetuses were examined by prenatal sonography. It was found that the overall prevalence of major anomalies among fetuses with non-consanguineous parents was 2.9% vs 10.9% for consanguineous parents. Thus, it is important that such information is made available in genetic counseling programs, especially in multi-ethnic and multi-religious communities, to enable couples to make informed decisions (Becker et al., 2015).

In December 1961, the World Health Organization (WHO) called a meeting of an Expert Committee on Human Genetics to prepare a report on the Teaching of Genetics in the Undergraduate Medical Curriculum and in Postgraduate Training (Ballantyne et al., 2006). The report can be obtained from Columbia University Press in New York, or the WHO in Geneva. Although the strength of the recommendations of the report was modified somewhat in view of the varying economic and political conditions in different parts of the world, it will be good ammunition for those of us who think genetics doesn't occupy enough space in our medical school curricula. After an introductory section on the value of genetics in the medical sciences, the report discusses undergraduate and postgraduate training. The genetic professionals are found in universities and medical centers and clinics. Recently, as the science of genetics has widened greatly, genetic professionals have grown more specialized. Therefore, they may be specialized in a particular disease i.e., cancer genetics, an age group i.e., adolescents, or a type of counseling i.e., prenatal. Every patient's case could be unique, therefore, the health care provider may refer the patient to a geneticist - a medical doctor or medical researcher - who specializes in that particular disease or disorder. The medical geneticist should be qualified enough as he/she must has completed a fellowship or has other advanced training in medical genetics. While a genetic counselor or genetic nurse may provide help to the patient with testing decisions and support issues, a medical geneticist will make the actual diagnosis of a disease or condition. There are many rare genetic diseases that only a geneticist can provide the most complete and current information about that particular condition (Cordier et al., 2016). The patient could be referred to a physician who is a specialist in the particular type of disorder the patient has, in order to assist the medical geneticist (Kinchen et al., 2004). Thus, if a patient is has a positive test for colon cancer, then the patient might be referred to an oncologist. Furthermore, for a diagnosis of

Huntington disease, the patient might be referred to a neurologist (New York-Mid-Atlantic Consortium for Genetic and Newborn Screening Services, 2009).

## 2.2. Degrees of Utilization of Genetic Counseling Worldwide

Most of the studies on genetic counseling were performed in the developed countries and the western societies, with only a few studies conducted in Arab countries (Al-Gazali, 2005; Teebi, 2010). In Australia, a study was performed to examine utilization of genetic counseling after diagnosis of a birth defect in 2004, and trends in utilization from 1991 to 2004. It was found that the utilization of genetic services by those classified as in "high need" of genetic counseling rose from 39.7% to 48.4% in 1991 and 2004 respectively. The "High need" defects included all single gene and chromosomal disorders, neural tube defects, dwarfing conditions, and patterns of malformation associated with known syndromes (Glynn et al., 2009).

When, the family decides to make use of the genetic counseling services, this decision of utilization of the genetic counseling service is influenced by various factors. These factors may be divided into geographical, cultural, financial and knowledge-based (Glynn et al., 2012). It was found that the genetic counseling utilization by parents who had a child or fetus with congenital malformations was according to the distorted phenotype, severity of the condition, type of birth and viability (Isabella et al., 2003). Furthermore, a study in Hawaii showed that genetic counseling facility utilization rates were much higher with the presence of multiple major birth defects, chromosomal abnormalities and malformation syndromes, certain specific birth defects, death of the fetus or infant and older maternal age (Forrester and Merz, 2007). Another study was conducted to report on prenatal utilization of ultrasound (US), amniocentesis (AC), and chorionic villus sampling (CVS) in pregnancies affected by birth defects in Hawaii. It was concluded that only a fraction of the Hawaii birth defects cases was prenatally diagnosed, additionally, the rates for prenatal US, AC/CVS, and prenatal diagnosis among pregnancies affected by birth defects were higher in 1998-2002 than in 1986-1991. AC/CVS rates were lower for maternal age <35 years (Forrester and Merz, 2006).

# 3. GENETIC DIAGNOSIS AND COUNSELING IN THE ARAB WORLD

## 3.1. View of Islam on Assisted Reproduction and Genetic Diagnosis

When dealing with the issue of assisted reproduction, Islamic teachings do not prohibit in vitro fertilization as long as the egg, sperm and uterus, are from a couple during the existence of a matrimonial bond. On the other hand, if divorce or death of a spouse occurs the use of any stored gametes or embryos, or a surrogate, is not permissible. All forms of reproductive technology that involve a donor egg or sperm are outlawed in Islam (Albar, 1999).

When Pre-implantation Genetic Diagnosis (PGD) technology first became available in the Middle East it was assumed by many policy makers that it would be preferred by all Muslim couples over prenatal diagnosis as it bypassed the issue of pregnancy termination. However, when couples at increased risk for a genetic disorder were advised of the technical aspects of PGD, including ovarian stimulation, oocyte retrieval, and the in vitro process, many opted to pursue prenatal testing with the option of termination of an affected pregnancy. In a study conducted in Saudi Arabia, 38% of couples expressed an interest in PGD – however acceptance rates varied significantly based upon which condition was discussed. Only 3 of 11 parents of children with cystic fibrosis (CF) were interested in PGD, whereas all seven parents of a child with thalassemia expressed an interest (Alsulaiman and Hewison, 2006). Of interest, individuals with graduate and postgraduate education were observed to have fewer concerns about PGD than those with less or no education. This may be due to better-educated parents having more knowledge about new reproductive technology and also having a better understanding of the Islamic law, which allows these procedures in certain circumstances. In summary, as is the case in the West, the acceptability of reproductive technologies must be established on an individual basis (Alsulaıman and Hewison, 2006).

## 3.2. Genetic Counseling and Genetic Disorders in Islamic Arab Countries

It should be noted that, while the term "Arab" includes individuals who are Christians, Jews, Druze and other minorities, the vast majority of Arabs are Muslims (Teebi and Teebi, 2005). Given that genetic counselors strive to respect the religious beliefs of the individuals and families they work with, when practicing in the Middle East, a genetic counselor should be aware of Islamic teachings and "fatwas," in order to best meet the needs of their patient population. A "fatwa" is an Islamic religious ruling or scholarly opinion on a matter of Islamic law. For rulings pertaining to medicine, the consensus group that develops the "fatwa" typically includes a broad and diverse representation of Islamic jurists and specialists, including clinicians and scientists from relevant disciplines, with the latter being responsible for providing the necessary background information (Al-Aqeel, 2007).

In Islam, bioethical decision making is carried out within the framework of the teachings of the "Holy Qura'n" and "Sunnah" which is the (statements, actions and approvals) of the Prophet Mohammed (peace be upon him) – and therefore are based on the interpretation of "shari'a" - Islamic Law (Al-Aqeel, 2007). A number of "fatwas" have been issued that have direct relevance to genetic counseling.

Genetic diseases, especially autosomal recessive diseases, are rare in the general population. However, they become unusually frequent in certain communities worldwide as a result of genetic isolation due to social, geographic or religious factors. When a new mutation is inserted in such a population it spreads rapidly, leading to an increased prevalence of carriers and a large number of affected homozygous individuals. There are high frequencies of genetic diseases and congenital disorders among Arab populations (Tadmouri et al., 2009; Teebi and Teebi, 2005). The high rates of genetic diseases and congenital disorders in the Arab populations can be attributed to several factors including: a) The high rate of traditional consanguineous marriages, which increases the frequency of autosomal recessive diseases; b) a relatively high birth rate of infants with chromosomal disorders related to advanced maternal age such as Down Syndrome and other trisomies; c) a relatively high birth rate of infants with malformations due to new dominant mutations related to advanced paternal age; d) large family sizes, which may increase the number of affected children in families with autosomal recessive conditions; and e) the lack of public health measures directed at the prevention of congenital and genetically determined disorders, and the shortage of genetic services and inadequate health care prior to and during pregnancy (Al-Gazali et al., 2006; Teebi, 2010). Thus, in Arab populations, several recessive diseases like Cystic Fibrosis, Phenylketonuria, Beta Thalassemia and Wilson disease are very frequent. However, due to the tribal society style life among Arab and consecutive genetic founder effects, the distribution of autosomal recessive disorders among the Arab populations is not uniformly distributed, but shows large geographic differences (Teebi and Teebi, 2005; Zolotogora, 2002).

## 3.3. The Genetic Counseling as a Profession

Many countries in the Middle East will report that genetic counseling services are currently available through their health care systems. It is important to note that although these countries may offer genetic counseling services via their clinical geneticists, primary health care providers and/or nurses – as of 2008, the Kingdom of Saudi Arabia and the United Arab Emirates are the only countries in the Middle East to have trained genetic counselors in practice. In terms of training programs, The Kingdom of Saudi Arabia is the only country in the Middle East to have had a formal genetic counseling training program. This training program was established in 2004 – as a 2.5 year post-graduate diploma at King Faisal Specialist Hospital and Research Centre in Riyadh. Currently, the genetic counselors practicing in the Kingdom of Saudi Arabia are a mixture of individuals trained locally in Riyadh, and Saudi nationals who returned to the Kingdom after training overseas. It was noted that genetic disorders with an autosomal recessive, inheritance pattern and the high rate of consanguineous marriages in Saudi Arabia are considered important contributors to neural tube defect (NTD) in this country (Seidahmed et al., 2014).

Most of the studies focusing on genetic counseling were done in western societies (Schoeffel et al., 2018), with only few studies conducted in Arab societies (Al-Gazali, 2005; Teebi, 2010). The medical genetic characteristics of these societies make genetic evaluation and counseling of great need. In addition, it is of utmost importance that counselees will comply with the recommendations given, in order to reduce the rate of morbidity, mortality and neonatal malformations, which is relatively high in the Arab society.

A study in Saudi Arabia, concluded that, despite NTD being a multifactorial pathogenetic disorder, there is a considerably low level of awareness in mothers of Spina bifida patients despite prevalence of this anomaly in eastern province. This necessitates an effort from health care providers to educate the community about such a birth defect entity, possible risk factors and preventable measures during pregnancy in addition to the folic acid supplement ahead of pregnancy planning and during the first 3 months of pregnancy. Furthermore, genetic counseling should be encouraged especially in those who have a positive familial history for better understanding (Othman et al., 2019).

Sickle cell disease (SCD) is an inherited autosomal recessive blood disorder, its high prevalence in Saudi Arabia is due to the high occurrence of consanguinity between first cousins (>50% of total marriages) (Jastaniah, 2011; Al Arrayed and Al Hajeri, 2010) and the population's lack of awareness of inherited hematological diseases (Al-Farsi et al., 2014). Previous studies have proven that, despite the legal implementation of compulsory premarital genetic counseling (PMGC), the incidence of SCD in Saudi Arabia has not changed significantly over the last 15 years (Ahmed et al., 2015; El-Hazmi, 2004; Alswaidi et al., 2012).

Currently there is a lack of genetic counseling training programs in the Middle East. As a result, the majority of individuals interested in this career must pursue their training overseas. Jordan currently has two nationals completing their training at Sarah Lawrence College in New York. These individuals returned to Jordan in 2009 where they began to establish the profession of genetic counseling in their country. It is anticipated that additional training programs in genetic counseling will be started in the Middle East in the next few years to meet the health care needs of these populations. Although it is possible for individuals from the Middle East to obtain training as a genetic counselor overseas, this is not ideal for a number of reasons. These new genetic counselors often return to completely different health care systems from the one in which they trained – in terms of resources, standards of practice, and a lack of recognition of the role of a genetic counselor in the provision of patient care. In addition, they frequently return to a country with vastly different cultural and religious norms, and significantly different prevalence rates for genetic disorders. As a result, there is a growing recognition amongst the genetic counseling community of the potential benefits of training genetic counselors in the country, or at least region, where they would like to practice. In the past, some countries in the Middle East have employed genetic counselors who are foreigners, typically from the West, to provide patient care. Again, this is not ideal for a number of reasons. Utilization of a genetic counselor who is a foreigner raises the probability that a patient will mask their true feelings due to concerns of how they will be perceived by an outsider, increases the potential for a misunderstanding of cultural norms and hampers effective communication secondary to the use of a translator (Panter-Brick, 1991). However, until a sufficient number of individuals from the Middle East can be trained as genetic counselors, hospitals will likely continue to utilize genetic counselors who are foreigners.

It is noteworthy that there is no a clear cut definition of genetic counseling that will apply to all those who practice this profession – as a definition inherently reflects the values, ethics, goals and medical practices of the person or group defining the practice (Resta, 2006). This is a critical factor for those practicing genetic counseling in the Middle East to consider, as the majority of the definitions of genetic counseling have been developed in North America, the United Kingdom, and Australia. As the profession of genetic counseling expands in the Middle East, a definition, or possibly definitions, reflective of practice in this region will likely be developed. In a study conducted in the Gaza Strip of Palestine found that though there is a decline in the consanguinity profile in the present compared to previous generation, consanguineous marriages are still a common practice there. This rationalizes the necessity for more awareness and counseling efforts about the potential health-related risks of consanguinity on individual lives and the population overall (Sirdah, 2014).

## 4. GENETIC COUNSELING IN ISRAEL

### 4.1. Introductory Background

When the State of Israel was created in 1948 the majority of Palestinian Arabs were forced out of their homeland to live either in Gaza Strip, the West Bank or surrounding Arab countries (Zlotogora, 2002). Those who remained in Israel at that time numbered about 156,000, but currently there are in excess of 1.6 million Arabs living in the country (Central Bureau of Statistics, Israel 2013; 2014). The Palestinians who remained in their homeland were later named by different names such as: Israeli Arabs, 1948 Arabs, Israeli Palestinians or Arab community in Israel. Most of the Arab population in Israel live in predominantly Arab cities and small towns or in mixed cities (Arabs and Jews) which are close to Jewish cities and towns, therefore, they have high exposure to westernized life style (Yitzhaki, 2010).

Israel is considered to be a modern, industrialized, multi-ethnic state, currently classified as a high-income country. Israel's high Human Development Index reflects achievements in key dimensions of human development, namely a long and healthy life, high educational attainment, and a decent standard of living (Sagar and Najam, 1998). On the other hand, Israel's high Gini index reflects high-income inequality (Yitzhaki, 1983). The population has grown rapidly as a result of both immigration and high fertility, and presently constitutes about 8.5 million people (Central Bureau of Statistics, Israel 2016). The two main population groups are Jews (74.8%) and Arabs (20.8%). Israeli Jews are customarily classified by regions of origin: Europe, America, Asia, North Africa, and native-born. Israeli Arabs include Muslims, Druze, and Christians. Between and within these groups, wide cultural, lifestyle, and socioeconomic variations exist. 12% of Jews were aged 65 years or older compared with 4% of Arabs (Central Bureau of Statistics, Israel 2016; Dwolatzky et al., 2017).

A substantially higher proportion of Arabs live in areas peripheral to the main urban centers, mainly in the northern and southern parts of Israel. The Bedouin Arabs are a distinct Muslim subgroup (about 13.4% of Israeli Arabs), most of whom reside in the southern part of the country. About half of them live in unauthorized villages, without access to water in their homes and sanitation infrastructure, but with access to health services. In general, important determinants of health inequalities have been shown to be income, education, sex, and ethnicity (Marmot, 2005; Link and Phelan, 1995; Blane, 1995; Cassel, 1976). The association between socioeconomic position and mortality has been suggested to be mediated through resources such as money, knowledge, prestige, power, and beneficial social connections that protect health, regardless of the specific mechanisms operating at any time (Link and Phelan 1995). Additional determinants of health inequalities include biological variation, free choice (i.e., behaviors), and environmental conditions that are mainly beyond the control of the individual and that characterize the conditions in which people are born, grow, live, work, and age (Marmot 2005; Link and Phelan 1995; Blane 1995; Cassel 1976).

In Israel, despite an overall increase in educational and economic status, the average educational attainment and income of Arabs remains below that of Jews. Additionally, participation of Arab women in the labor force is around 25%, compared with 70% among Jewish women (Central Bureau of Statistics, Israel 2016). In 2014, 22% of the Israeli population lived under the national poverty line, the second highest among the Organization for Economic Co-operation and Development (OECD) countries. Poverty is particularly high among ultra-Orthodox Israeli Jews (59%) and Israeli Arabs (54%). Inequality levels, measured by the Gini coefficient (in which higher values reflect greater inequality), increased from 0.353 in 1998, to 0.371 in 2014, fourth in rank among OECD countries (OECD average 0.308) (National Insurance Institute, Israel 2015).

All permanent residents, in Israel, are medically insured under the National Health Insurance Law and are members of one of the four Health Ministry Organizations (HMOs) that supply health services in the community that are included in a nationally determined basket of services. All HMOs support and cooperate with the program, including in the development, assessment and publication of quality indicators (Gross et al., 2001).

## 4.2. Ethnic Communities in Israel

The Israeli population includes many different subpopulations (Jews of various origins, Arab Muslims, Arab Christians, Druze, Bedouins, and so on). The Jews in the Israeli population were divided into isolated ethnic groups that were separated geographically throughout most of the other countries of the world. Within these semi-isolated ethnic groups, the need and desire to maintain religion and culture led to relatively high levels of intra-community marriages. Similarly, individuals of Arab-Muslim origin tend to live in villages that were founded a few generations ago by a small number of individuals. In accordance with the customs of the Arab population throughout the Middle East, consanguineous marriages are common among Israeli Arabs, with a preference for first-cousin marriages. The Bedouin population is unique in its structure, which is tribal, and exhibits an extremely high level of consanguinity (Hanany et al., 2018).

Therefore, it is a known fact that Israel is composed of many ethnic groups. The Jewish population of Israel counted about 5 Millions in 2016 (Central Bureau of Statistics, Israel, 2016), who are composed of three major communities: Sephardim and Oriental Jews (about 53% percent) and Ashkenazim (about 47% percent). These communities, with their subgroups, descended from tens of other countries. In many instances each major community, along with its component groups, can be characterized genetically by the high frequency of certain genetic disorders found within these varying Jewish ethnic divisions (Goodman, 1979). An autosomal recessive genetic disorder known as Gaucher disease (GD), is the most common genetic disorder in Ashkenazi Jews (Alcalay et al., 2014). However, screening of (GD) is offered to Ashkenazi Jews worldwide and has been offered in Israel since 1995. The carrier screening procedure is performed in order to prevent severe, untreatable genetic disease by identifying couples at risk before the birth of an affected child, and providing such couples with options for reproductive outcomes for affected pregnancies (Zuckerman et al., 2007).

Among the Muslim Arabs, for example, the Bedouin represent a distinct community, living as an isolated population for centuries. This society is traditional, exhibiting polygamy, high birth rates, and among the highest rates of consanguinity, up to 60% (Vardi-Saliternik et al., 2002). Thus, it is subject to two well-known phenomena in population genetics – the Founder effect (when a specific pathogenic genetic variant can be traced back to few original founders), and the Wahlund effect (exhibiting a decreased number of heterozygotes compared to expected by Hardy-Weinberg equilibrium, caused by population stratification) (Shalev, 2019). In such populations, homozygosity for a pathogenic genetic variant can be diagnosed as early as three generations after its appearance in the first carrier (Zlotogora et al., 2007).

All the above points led to a higher prevalence of autosomal recessive diseases, in the Israeli population compared to other populations around the world. Genetic analysis of these inherited diseases in the different sub-populations led to the discovery and characterization of founder mutations. The development of new sequencing strategies known as next-generation sequencing (NGS) revolutionized genetic studies, mainly due to the high amount of sequencing data generated by these methods. Subsequently, a large number of highly important NGS-based databases is now available, including gnomAD (Lek et al., 2016), as well as those linking specific mutations to Mendelian diseases, including ClinVar (Landrum et al., 2015), Leiden open variation database (LOVD) and the human gene mutation database (HGMD). These databases provide an excellent tool for studying the pattern and frequency of variants and diseases in different populations. Thus, in Israel it becomes illogical to ignore the relevance of medical genetics among the various ethnic groups consisting the entire population (Fokkema et al., 2011; Cremers et al., 2014; Cornelis et al., 2017).

One of our previous studies was conducted for screening of Familial Mediterranean Fever- FMF,(known as *MEFV* gene) gene mutations (which is an autosomal recessive genetic disease) in the major ethnic communities of Israel *viz,* Jews (Ashkenazi and non-Ashkenazi), Arabs (Muslims and Christians) and Druze. Our study comprehensively provided a spectrum of FMF mutations in various communities of Israeli society (Sharkia et al., 2013). Furthermore, a recent report for us described the main clinical features of cerebro-facio-thoracic dysplasia (CFTD) syndrome, which was found to be caused by a mutation in *TMCO1* gene that is inherited by an autosomal recessive manner. This gene was found in two members of a consanguineous family from an Arab community in Israel (Sharkia et al., 2019). In one of our studies we had a case report, in which we investigated - clinically and genetically, two patients (sisters) from a consanguineous Palestinian family from the Israeli Arab community. This study was considered to be the first report for Sanfilippo syndrome type A, which is an autosomal recessive lysosomal storage disorder in Israel (Sharkia et al., 2014). We had further clinical and genetic investigations of various mutations in consanguineous families of the Israeli Arab community. These genetic mutations include; *PTRH2* gene and *ADAT3* gene (Sharkia et al., 2017; Sharkia et al., 2018).

## 4.3. Characteristics of the Arab Society in Israel

The Arab population in Israel, which today counts about 1.8 million (Central Bureau of Statistics, Israel 2016) is an ethnic group with some unique cultural, religious and social characteristics, which differ from those of the general population in Israel. This population belongs to various religions: Muslims (83.9%), Christians (8.3%) and Druze (7.8%).

The Arab communities are still undergoing a transition from a mainly agrarian society, to a more urbanized one. On the whole, compared to the Jewish population, Arabs in Israel have lower socioeconomic status and poorer health awareness (Daoud, 2008). The community is characterized by a high rate of consanguineous marriages with a common founder effect (Central Bureau of Statistics, Israel 2016; Sharkia et al., 2016). Recently, a westernized lifestyle was adopted by most of the Arab population of Israel (Treister-Goltzman and Peleg, 2015). It was found that the main causes of death that might contribute to the lower age of Arabs than Jews could be due to chronic diseases; especially ischemic heart disease and diabetes (Na'amnih et al., 2010). Recently, our previous study found that the incidence rate of type 2 diabetes was increasing significantly from 11.3% to 17.7% in the years 2005 and 2015 respectively, with a progressive increase with age in both genders (Sharkia et al., 2018).

## 4.4. Medical Genetics Education in Israel

Concerning postgraduate medical training in medical genetics, no official programs are offered in Israel because the Israeli Board of Medical Specialties does not recognize the specialty of medical genetics. Two universities (Tel-Aviv and Hebrew universities) offer advanced degrees

(MS and PhD) in human genetics, but only the Hebrew University at present awards the MS degree in genetic counseling. Medical genetics is taught as a distinct course in two of the four medical schools: Tel-Aviv and Jerusalem, both of which have departments of human genetics. In the other two medical schools (Haifa and Beersheva), matters pertaining to medical genetics are dispersed throughout other courses (Goodman, 1984).

This scattered educational exposure to the teaching of genetics in medicine is not unique to Israel only but, similar pattern also exists in the United States and in most other parts of the world (Childs et al., 1981). As a result of these practices, many practitioners of medicine view medical genetics as a secondary subject not relevant to the more common problems in medicine. This restricted view limits the capabilities of medical genetics and also diminishes the scope of care given to many patients and families who need them.

## 4.5. Centers of Medical Genetics in Israel

In Israel, all centers for the practice of medical genetics are located in major teaching hospitals. Jerusalem has one center; the Tel-Aviv area, five; Rehovot, one; Haifa, one; and Beersheva, one. Referrals to genetic centers come both from within the hospital setting and from the hundreds of medical outpatient clinics located throughout the country. For the most part these outpatient clinics are staffed by physicians who did not receive their medical education in Israel but were trained in countries where little or no emphasis was given to the teaching of medical genetics (Goodman, 1984). In the year 2003, there were 12 medical genetic centers (Zuckerman et al., 2007). But recently, in the year 2019 the number of centers, dealing with medical genetics testing, according to the information available from the website of the ministry of health (MOH) here in Israel, has reached to 19 centers (Ministry of Health, Israel, 2019).

## 4.6. Hardships for Genetic Investigations and Counseling

A barrier that hinder or even hardens the process of genetic investigation and counseling is caused by an increased risk for co-inheritance of two or even more rare recessive disorders (Falik-Zaccai et al., 2008; Shalev, 2019). Furthermore, different conditions can present with very similar phenotypes within the same family. This results in unique complications of the consultation that require extra caution and thoughtfulness from the genetic counselor. In addition, Bedouin society in general has low socio-economic and educational levels (Treister-Goltzman & Peleg, 2014), which contribute to the decreased awareness of the possibility of genetic disorders in this population, and frequent difficulties in supplying the relevant documentation and in understanding the messages of genetic counseling. Also, a possible tendency to obscure genetic disorders may exist, fearing that it might stigmatize the family and interfere with marriage. Prenatal counseling imposes further complications, resulting from reduced acceptance of the genetic diagnosis and Muslim ban on pregnancy termination after specific dates of pregnancy (Jaber et al., 2000; Raz & Atar, 2004).

## 4.7. Degrees of Utilization of Genetic Counseling in Israel

In a multi-cultural country like Israel, various responses from the concerned family regarding the factors that may influence the utilization of genetic counseling services, that is dependent on personal and family backgrounds. Genetic counseling service has been available in Israel since 1991, at the genetic counseling units in medical centers and partly at community genetics clinics (Zlotogora et al., 2006). The Arab population in Israel counts about 1.8 million (Central Bureau of Statistics, Israel 2016). The Arab community in Israel is characterized by some unique features *viz*, most of the populations are Muslims and live in villages and towns, which were founded by small numbers of individuals. They have high fertility rates and a high rate of consanguineous marriages (Sharkia et al., 2008; Vardi-Saliternik et al., 2002). They also have underutilization of prenatal diagnosis services and low rate of pregnancy termination of an affected fetus, these are some of the risk factors leading to high prevalence of infant morbidity and mortality in the Israeli Arab community (Muhsen et al., 2010; Sharkia et al., 2010; Tarabeia et al., 2004; Zlotogora, 2002). In one of our research articles we discussed the various factors affecting the utilization of genetic counseling services among 414 pregnant Israeli Arab women. We found that the underutilization of genetic counseling services among pregnant Israeli Arab women was associated with lower income level, attitude toward genetic counseling, accessibility to service and religiosity. The study recommended the proper expansion of genetic counseling service within this community of the Israeli society (Sharkia et al., 2015).

## 5. IMPROVEMENT OF GENETIC COUNSELING IN ISRAEL

From the above mentioned review on genetic counseling worldwide, in the developing countries, in the Arab countries and locally here in Israel, the importance of genetic counseling has become obvious not only for the current population, but for the future generations also. Therefore, there are various suggestions that could effectively improve upon genetic counseling in Israel. These suggestions include the following points:

1. Establishment of a National Registry for Genetic Diseases. Such a registry would not only tremendously aid in the counseling and care being given but would advance the research efforts pertaining to the recognition of new genetic diseases and those specific disorders involving certain local ethnic communities.
2. Creation of a computer center for the storage, retrieval, and analysis of all information regarding genetic counseling in Israel.
3. Initiation of national periodic review conferences to discuss ways of improving all aspects of genetic services, including that of counseling.

4. Publication of a quarterly bulletin or newsletter to keep physicians abreast of the rapid progress in genetics and the applicability of genetics in medicine.
5. Physical improvement of outpatient facilities for genetic patients.
6. Involvement of psychiatrists and psychologists in genetic counseling to help us better understand our patients.
7. An increase in the number of university training programs for those who wish to become genetic counselors (Goodman, 1984).

Therefore, Genetic counseling can be enhanced for the entire population of each country particularly, those countries comprised of many ethnic groups, such as Israel, and of special concern for the families at high risk who are faced with genetic disorders by using proper interventions for a better decision making of the family.

# 6. OUR EXPERIENCE, RECOMMENDATIONS AND SUGGESTIONS

## 6.1. General Background

Genetic counseling during pregnancy can lead to meaningful decisions that may have long-term influence on the family, mainly in the sense of family planning. The most prevalent step that can be taken presently in order to avoid the birth of affected child is the termination of abnormal pregnancy. Aborting a genetically affected fetus is a decision that should be taken solely by the family itself, without any external pressures. Thus, the purpose of genetic counseling services is to allow the couples to take the proper decision regarding their pregnancy, after discussion and providing them with the required information about various diagnostic tests and preventive measures, which is expected to decrease congenital morbidity rate.

## 6.2. Education and Awareness Programs

For the successful implementation of any prevention program, adequate awareness, in the community, is essential. 'Genetic services' touch a wider group than ordinary health services, since genetic information affects an entire family and even the relatives and not just the individual patient who is really the affected person. It may be predictive of future adverse events in an individual or family members' health, and the choices made, based on the genetic information, may affect future generations. Thus, careful understanding of the defect, its genetic basis, its inheritance pattern, its clinical consequences, mode of treatment, and ways and means of prevention is necessary. This requires education and awareness as well as adoption of preventive measures, including pre-marital screening and early diagnosis and intervention, including inborn screening. Unless the general population is educated and made aware of the prevalent conditions, steps towards control and prevention will be inefficient.

In Islamic/Arab communities, overall literacy is still lower than in developed countries. However, a sufficient amount of information may be conveyed to affected persons, their families and the overall population by a well-designed awareness programs, involving preparation of booklets in Arabic, posters, lectures, videos, articles in newspapers, inclusion of biology curricula in secondary schools, and doctor-patient, doctor-family, doctor-doctor and family-family interactions.

It is now obvious that genetic counseling is a component of several medical disciplines, but genetic counseling plays an essential role in control, prevention, or even minimizing genetic disorders and is considered inseparable from genetic diagnosis. Its role in medical genetics is of particular significance as it utilizes the predictive nature of genetic decisions to estimate the risk of recurrence in other members of the family, as well as to help increase the knowledge of the family about the causes, the diagnostic aspects, the available treatment strategies and ways and means of achieving the ultimate goal of prevention. An important aspect of genetic counseling is the provision of the estimated recurrence risk to the families with a genetic disease, patients or other members of the family, or carrier individuals planning to get married or conceive, or individuals who have been exposed to harmful environmental factors.

Therefore, it is important to note that genetic counseling deals with three major elements to fulfill its objectives i.e., control and prevention, these elements are: (i) diagnostic aspects, where an accurate diagnosis is required for a secure foundation of the advice, (ii) the calculation of recurrence risk, and (iii) a communicative role, providing beneficial information about the disease and available supportive, preventive and ameliorative measures, to ensure that those who are advised will benefit.

In light of the previously mentioned points, it becomes of crucial importance to utilize the process of genetic counseling in order to decrease the frequency of the genetic diseases, inborn deformities and infants' mortality. Therefore, educational and awareness programs are very important in order to spread the knowledge among the population about the drawbacks of genetic disorders and consanguinity and their effect on the family as well as on the entire society.

## 6.3. Our Personal Experience in Our Local Society

In one of our unpublished research work, we conducted a study in order to determine the knowledge level and attitudes of class-X students towards genetic disorders, consanguinity and genetic counseling. For this purpose, two rural populated areas from the northern triangle region, of the Arab society in Israel, were chosen for implementation of this research work. This choice was based on differences in the socio-economic standard of the two villages. The research tool consisted of two questionnaires; the first one was knowledge and attitudes questionnaire before implementing the program and the second one was knowledge and attitudes evaluation questionnaire after implementing the program. As a part of the whole project, an educative and preventive program was implemented in the two villages selected. The effect of this program on the students' knowledge level and attitudes was evaluated. It was possible to find a distinguished co-relation between the socio-economic standard of the village and the students' knowledge levels. About 95% of the students from village "A" (the village with a low socio-economic standard), had medium and low knowledge levels, and only 5% of them had high knowledge levels. On the other hand, about 75% of the students from village "B" (the village with a high socio-economic standard), had medium and low knowledge levels, and 25% of them had high knowledge levels. It is also clearly noticed that, in village "B" 11% of the students had low knowledge levels, while that in village "A" it is as high as 50%. The relationship between the socio-economic standard of the village and the students' attitudes was examined, but no appreciable difference could be found. A total of 118 students (84%) opposed consanguineous marriages and 109 students (75%) supported genetic counseling and undergoing genetic medical examination before and after marriage. The results indicated that some factors do affect the students' attitudes. There was no clear co-relation between the parent's consanguineous marriages and the student's attitude. On the other hand, the number of the students' family members constituted a true reflection to the variability of their attitudes. The majority of the students who supported consanguinity (77%) and opposed undergoing genetic medical examination for genetic counseling (71%) belong to families that have more than 5 members. The results showed that the percentage of those who oppose consanguinity and those who support genetic counseling were similarly distributed among the various knowledge levels. Furthermore, the results indicated the absence of a relationship between the students' knowledge level in consanguinity, genetic disorders and genetic counseling and their attitudes. We could conclude that the results obtained from this research work emphasizes the need to change and improve the students' knowledge about genetic disorders, consanguinity and genetic counseling. The results indicate that the increase in the students' knowledge level leads to a positive feedback in their attitudes. These facts suggest that the students' preliminary attitudes were not based on correct scientific knowledge. Therefore, there is an urgent need for the students to

**Table 1. Summary of some genetic counseling studies in Israel**

| Number | Reference/Year | Subjects | Study results |
|---|---|---|---|
| 1 | Bakst et al., 2019 | One thousand and sixty-seven eligible Israeli female subjects were recruited for the study, within 8–72 hours after the delivery of a live born infant. Nine hundred and ninety-six women were ultimately allocated; of whom 766 agreed to interview (76.9%) and 228 (21.4%) refused. Nearly, three quarters of the subjects were Jewish (N¼569, 74.3%) as compared with about a quarter of Arabs (N¼197, 25.7%). | Jews had significantly higher proportions on the following background factors as compared with Arabs: they were more likely to be older, have more than 12 years of education (among both subject and spouse), self-identified as secular, have a higher monthly income and have auxiliary medical insurance. Jewish women who underwent triple test screening were (in descending order): more accepting of the notion of abortion, less religious, had supplementary insurance, fewer children, higher income, positive views on a screening, and were younger as compared with those that declined testing. Analysis for the Arab participants showed that having supplementary medical insurance ($p < .01$), a higher monthly income ($p < .01$) and fewer children ($p < .05$) was significantly associated with the triple test uptake. It could be concluded that this study showed that background influences and having supportive attitudes on screening and abortion are for the most part, related to prenatal test use; however, these often involve dense decision-making practices that can be difficult to approximate. Hence, provision of ethical prenatal care during pregnancy that would routinely evaluate individual preferences while communicating the practical aspects of genetic testing alongside accurate estimates-of-risk could allow women to make truly informed and autonomous prenatal test choices. This can, in turn, provide a better approximation of the social need for prenatal screening in Israel. |
| 2 | Bernstein-Molho et al., 2019 | Consecutive Israeli Arab breast and/or ovarian cancer patients (a total of188 patients) were recruited. Breast cancer patients were offered BRCA sequencing and deletion/duplication analysis after genetic counseling. | Overall, 150 breast cancer cases (median age at diagnosis: 40 years, range 22-67) and 38 had ovarian cancer (median age at diagnosis: 52.5 years, range 26-79). The overall yield of comprehensive BRCA1/2 testing in high-risk Israeli Arab individuals is low in breast cancer patients, and much higher in ovarian cancer patients. The results may guide optimal cancer susceptibility testing strategy in the Arab-Israeli population. |

| Number | Reference/Year | Subjects | Study results |
|---|---|---|---|
| 3 | Eyal, 2019. | All cases of de novo mutations from a pool of 2,260 pregnancies for which prenatal molecular diagnosis was applied, between the years 2008 and 2017, were sorted in a tertiary center in Israel, over a 10-year period. A total of 122 molecular prenatal diagnosis performed for de novo mutations, in 90 women, were identified | There was a large increase in the annual number of prenatal diagnoses performed due to a previous pregnancy with a de novo mutation. This reflected the growing understanding regarding the role of these mutations in the pathogenesis of genetic diseases. |
| 4 | Sharony et al., 2018 | A cohort of 181 women with known pregnancy course and neonatal follow-up data was studied. They had genetic counseling during the third trimester, a known pregnancy course, and neonatal follow-up data. | Fifty-two patients (group 1—29%) met criteria for termination of pregnancy, and 129 (group 2—71%) were not eligible at the time they were referred for genetic counseling (GC). The GC resulted in 104 (group 3—57%) who followed all the recommendations such as targeted US, fetal MRI, and amniocentesis and 77 women who declined amniocentesis (group 4—43%). |
| 5 | Gesser-Edelsburg and Shahbari, 2017 | The study included 29 participants: 24 patients and 5 doctors | The results showed that the emotional element is no less dominant than religious and social elements. The findings emphasized the disparities between doctors and women regarding emotional involvement (non-directive counseling). The women interviewees (N = 24) felt that this expressed insensitivity. The emotional component has not been raised in previous studies of Muslim women at high risk for congenital defects in their fetus, and therefore comprises a significant contribution of the present study. |
| 6 | Siani and Assaraf, 2017 | Qualitative data was collected from 15 semi-structured interviews with Israeli genetics experts. | The study recommended that the genetic counseling sessions should be lengthened so that counselors will be able to truly gauge all the prior knowledge of the counselees, their religious beliefs, norms, values and attitudes towards genetic testing. It also recommended that students continue to study genetics further into high school. It also suggested adding a preparation session, similar to a prenatal course, to the genetic counseling of lay people so that their genetic knowledge, attitudes and perceptions will be enhanced, leading to more efficient genetic counseling and more informed decisions. |

# Table 1. (Continued)

| Number | Reference/Year | Subjects | Study results |
|---|---|---|---|
| 7 | Sharkia et al., 2015 | A case–control study was conducted among 414 pregnant Arab women who were referred by a family physician or a perinatologist to genetic counseling services between 2008 and 2011. Data was collected using interviews, with both groups 'users' and 'non-users' of genetic counseling, based on a structured questionnaire including demographic, socioeconomic, medical and cultural variables. | In multivariate analysis, factors affecting women's utilization of genetic counseling service were high income level (OR 3.44, 95%CI 1.8–6.5, p < 0.001), high service accessibility (OR 0.75, 95%CI 0.67–0.84, p = 0.001), more positive attitude toward genetic counseling (OR 0.43, 95%CI 0.27–0.67, p = 0.012) and lower religiosity level (OR 1.40, 95%CI 0.94–2.09, p = 0.04). |
| 8 | Hashiloni -Dolev and Weiner, 2008 | This is a comparative study, which was conducted to evaluate the Israeli and German genetic counselors' perceptions of the moral standing of the fetus. Data collected through in-depth interviews with counselors in both countries (N = 32) are presented, and their moral practices are analyzed. | The findings suggest that while German counselors perceive the fetus as an autonomous being and debate the particular biological stages through which this autonomy is acquired; Israeli counselors do not consider the moral status of the fetus independently of its relations with its family hence, deploying a 'relational ethics.' |
| 9 | Sa'd et al., 2006 | The study assessed 55 Epidermolysis bullosa (EB) families, from Israel over a period of 5 years, for pathogenic sequence alterations in the 10 genes known to be associated with EB. | It was found that the molecular epidemiology of EB in the Middle East is significantly different from that previously delineated in Europe and the US. The obtained data raise the possibility that similar differences may also be found in other genetically heterogeneous groups of disorders, and indicate the need for population-specific diagnostic and management approaches. |
| 10 | Jaber et al., 2000 | A total of 231 Arab women of childbearing age were interviewed 3 days postpartum to assess their knowledge of prenatal diagnosis and termination of pregnancy, their willingness to undergo prenatal diagnosis, and their opinions on termination of pregnancy in the event of a severely affected fetus. | Half of the women believed that prenatal testing is not an effective (or accurate) tool for diagnosing an affected fetus. A quarter had poor knowledge on prenatal diagnosis, and a quarter believed that prenatal diagnosis does provide the correct diagnosis. 95% said they would agree to undergo prenatal diagnosis, and in the case of a severely affected fetus, 36% said they would agree to terminate the pregnancy, 57% said they would not, and 7% were undecided. |

| Number | Reference/Year | Subjects | Study results |
|---|---|---|---|
| 11 | Sheiner et al., 1998. | The study was designed to identify predictors of parental decision whether to terminate a pregnancy after a diagnosis of a major congenital malformation in a traditional society of the Bedouin Arabs in southern Israel. The data were abstracted from medical records of 295 families who sought counseling in the third level ultrasound clinic between 1990 and 1996. | The diagnosis of a major malformation was confirmed in 64% of the cases. Pregnancy termination was a realistic option for 125 women (66.5%) as the rest were too advanced in their pregnancy. Such a delay was less common in cases of multiple malformations than in a single malformation (19.2% versus 39.0% respectively). Forty-nine of the 125 women (39.2%) chose to terminate their pregnancy. The only significant predictors of termination decision were earlier gestational week at diagnosis and previous uncompleted pregnancies. The findings indicate the importance of promoting early genetic counseling and early prenatal diagnosis, for any population where abortions are not readily acceptable. |
| 12 | Chiba-Falek et al., 1998 | The screened population comprised 497 students from one school, which all the children of the village attend. | The results revealed high carrier frequency, 8.5%, for the two CFTR mutations, G85E and ¢F508, and a carrier frequency of 12% for the 5T allele. The screening results were reported to the physicians of the village to be used, upon request, for genetic counseling. |

understand and comprehend this subject in order to have a logical and confident futuristic attitudes. In order to improve the students' knowledge in these subjects, it is of crucial importance to implement certain preventive and educative programs to the secondary school students of the Arab society in Israel. Furthermore, this program aims at protecting the health of future generations from genetic disorders. Therefore, we believe that warning the students from the risks of consanguineous marriages leads to healthy future generations. Thus, the whole society avoids the economic burden and the psychological sufferings caused by genetic disorders and malformation.

# REFERENCES

Abacan, M., Alsubaie, L., Barlow-Stewart, K., Caanen, B., Cordier, C., Courtney, E., Davoine, E., Edwards, J., Elackatt, N. J., Gardiner, K. and Guan, Y., 2018. The global state of the genetic counseling profession. *European Journal of Human Genetics*, p. 1.

Ahmed, A. E., Alaskar, A. S., Al-Suliman, A. M., Jazieh, A. R., McClish, D. K., Al Salamah, M., Ali, Y. Z., Malhan, H., Mendoza, M. A., Gorashi, A. O. and El-Toum, M. E., 2015. Health-related quality of life in patients with sickle cell disease in Saudi Arabia. *Health and quality of life outcomes*, *13*(1), p. 183.

Al Aqeel, A. I., 2007. Islamic ethical framework for research into and prevention of genetic diseases. *Nature Genetics*, *39*(11), p. 1293.

Al Arrayed, S. and Al Hajeri, A., 2010. Public awareness of sickle cell disease in Bahrain. *Annals of Saudi medicine*, *30*(4), pp. 284-288.

Albar MA (1999) Counseling about genetic disease: an Islamic perspective. Easter Med Health J 5:1129–1133

Alcalay, R. N., Dinur, T., Quinn, T., Sakanaka, K., Levy, O., Waters, C., Fahn, S., Dorovski, T., Chung, W. K., Pauciulo, M. and Nichols, W., 2014. Comparison of Parkinson risk in Ashkenazi Jewish patients with Gaucher disease and GBA heterozygotes. *JAMA neurology*, *71*(6), pp. 752-757.

Al-Farsi, O. A., Al-Farsi, Y.M., Gupta, I., Ouhtit, A., Al-Farsi, K.S. and Al-Adawi, S., 2014. A study on knowledge, attitude, and practice towards premarital carrier screening among adults attending primary healthcare centers in a region in Oman. *BMC Public Health*, *14*(1), p. 380.

Al-Gazali, L.; Hamamy, H. and Al-Arrayad, S., 2009. Genetic disorders in Arab world. *British Medical Journal,* Vol. 333, pp. 831-843.

Al-Gazali, L. I., 2005. Attitudes toward genetic counseling in the United Arab Emirates. *Public Health Genomics*, *8*(1), pp. 48-51.

Alsulaiman A, Hewison J (2006) Attitudes to prenatal and preimplantation diagnosis in Saudi parents at genetic risk. Prenat Diagn 26:1010–1014.

Alswaidi, F. M., Memish, Z.A., O'Brien, S. J., Al-Hamdan, N. A., Al-Enzy, F. M., Alhayani, O.A. and Al-Wadey, A. M., 2012. At-risk marriages after compulsory premarital testing and counseling for β-thalassemia and sickle cell disease in Saudi Arabia, 2005–2006. *Journal of genetic counseling*, *21*(2), pp. 243-255.

Anderson, V. E., 2003. Sheldon C. Reed, PhD (November 7, 1910–February 1, 2003): Genetic Counseling, Behavioral Genetics. *The American Journal of Human Genetics*, *73*(1), pp. 1-4.

Bakst, S., Romano-Zelekha, O., Ostrovsky, J. and Shohat, T., 2019. Determinants associated with making prenatal screening decisions in a national study. *Journal of Obstetrics and Gynaecology*, *39*(1), pp. 41-48.

Ballantyne, A., Goold, I. and Pearn, A., 2006. *Medical genetic services in developing countries: the ethical, legal and social implications of genetic testing and screening.*

Bayrak-Toydemir P, Stevenson D. *Capillary Malformation-Arteriovenous Malformation Syndrome.* 2011 Feb 22 (Updated 2019 Sep 12). In: Adam MP, Ardinger HH, Pagon RA, et al., editors. GeneReviews® (Internet). Seattle (WA): University of Washington, Seattle; 1993-2019. Available from: https://www.ncbi.nlm.nih.gov/books/NBK52764/.

Becker, R., Keller, T., Wegner, R. D., Neitzel, H., Stumm, M., Knoll, U., Stärk, M., Fangerau, H. and Bittles, A., 2015. Consanguinity and pregnancy outcomes in a multi-ethnic, metropolitan European population. *Prenatal diagnosis*, *35*(1), pp. 81-89.

Bernstein-Molho, R., Barnes-Kedar, I., Ludman, M. D., Reznik, G., Feldman, H. B., Samra, N. N., Eilat, A., Peretz, T., Peretz, L. P., Shapira, T. and Magal, N., 2019. The yield of full BRCA1/2 genotyping in Israeli Arab high-risk breast/ovarian cancer patients. *Breast cancer research and treatment*, pp. 1-7.

Biesecker, B. B. and Peters, K. F., 2001. Process studies in genetic counseling: peering into the black box. *American journal of medical genetics*, *106*(3), pp. 191-198.

Blane, D., 1995. Social determinants of health--socioeconomic status, social class, and ethnicity. *American journal of public health*, *85*(7), pp. 903-905.

Brock, D. J. H., 1974. Prenatal diagnosis and genetic counseling. *Journal of Clinical Pathology. Supplement (Royal College of Pathologists)*. 8, p. 150.

Butler, J. M., 2007. Short tandem repeat typing technologies used in human identity testing. *Biotechniques*, *43*(4), pp. Sii-Sv.

Cassel, J., 1976. The contribution of the social environment to host resistance: the Fourth Wade Hampton Frost Lecture. *American journal of epidemiology*, *104*(2), pp. 107-123.

Caulfield LLM, T., 1998. The commercialization of human genetics: profits and problems. *Molecular medicine today*, *4*(4), pp. 148-150.

*Central Bureau of Statistics Israel (2014)*. Available from: www.cbs.gov.il. At: http://www.cbs.gov.il/shnaton65/st02_01.pdf).

Central Bureau of Statistics, Israel (2013). *65th Independence Day*. Available from http://www.cbs.gov.il/.

Central Bureau of Statistics, Israel (2014) *Statistical Abstract of Israel*. URL: http://www.cbs.gov.il/.

Central Bureau of Statistics, Israel (2016). *Statistical abstract of Israel*. 66 edn. Jerusalem: Israel Central Bureau of Statistics.

Chaabouni, H., Chaabouni, M., Maazoul, F., M'Rad, R., Jemaa, L. B., Smaoui, N., Terras, K., Kammoun, H., Belghith, N., Ridene, H. and Oueslati, B., 2001, April. Prenatal diagnosis of chromosome disorders in Tunisian population. In *Annales de genetique* (Vol. 44, No. 2, pp. 99-104). Elsevier Masson.

Chandler, N. J., Ahlfors, H., Drury, S., Mellis, R., Hill, M., McKay, F. J., Collinson, C., Hayward, J., Jenkins, L. and Chitty, L. S., 2019. Noninvasive Prenatal Diagnosis for Cystic Fibrosis: Implementation, Uptake, Outcome, and Implications. *Clinical Chemistry*, pp. clinchem-2019.

Chiba-Falek, O., Nissim-Rafinia, M., Argaman, Z., Genem, A., Moran, I., Kerem, E. and Kerem, B., 1998. Screening of CFTR mutations in an isolated population: identification of carriers and patients. *European Journal of Human Genetics*, 6(2), p. 181.

Childs, B., Huether, C. A. and Murphy, E. A., 1981. Human genetics teaching in US medical schools. *American journal of human genetics*, 33(1), p. 1.

Clark, J. R., Scott, S. D., Jack, A. L., Lee, H., Mason, J., Carter, G. I., Pearce, L., Jackson, T., Clouston, H., Sproul, A. and Keen, L., 2015. Monitoring of chimerism following allogeneic haematopoietic stem cell transplantation (HSCT): technical recommendations for the use of short tandem repeat (STR) based techniques, on behalf of the United Kingdom National External Quality Assessment Service for Leucocyte Immunophenotyping Chimerism Working Group. *British journal of haematology*, 168(1), pp. 26-37.

Cordier, C., Taris, N., Moldovan, R., Sobol, H. and Voelckel, M. A., 2016. Genetic professionals' views on genetic counsellors: a French survey. *Journal of community genetics*, 7(1), pp. 51-55.

Cornelis, S. S., Bax, N. M., Zernant, J., Allikmets, R., Fritsche, L. G., den Dunnen, J. T., Ajmal, M., Hoyng, C. B. and Cremers, F. P., 2017. In silico functional meta-analysis of 5,962 ABCA4 variants in 3,928 retinal dystrophy cases. *Human mutation*, 38(4), pp. 400-408.

Cremers, F. P., den Dunnen, J. T., Ajmal, M., Hussain, A., Preising, M. N., Daiger, S. P. and Qamar, R., 2014. Comprehensive registration of DNA sequence variants associated with inherited retinal diseases in Leiden Open Variation Databases. *Human mutation*, 35(1), p. 147.

Daoud, N., 2008. Challenges facing minority women in achieving good health: voices of Arab women in Israel. *Women & health*, *48*(2), pp. 145-166.

Drummond, C. L., Gomes, D. M., Senat, M. V., Audibert, F., Dorion, A. and Ville, Y., 2003. Fetal karyotyping after 28 weeks of gestation for late ultrasound findings in a low risk population. *Prenatal Diagnosis: Published in Affiliation With the International Society for Prenatal Diagnosis*, *23*(13), pp. 1068-1072.

Dwolatzky, T., Brodsky, J., Azaiza, F., Clarfield, A. M., Jacobs, J. M. and Litwin, H., 2017. Coming of age: health-care challenges of an ageing population in Israel. *The Lancet*, *389*(10088), pp. 2542-2550.

Eikmans, M., van Halteren, A. G., van Besien, K., van Rood, J. J., Drabbels, J. J. and Claas, F. H., 2014. Naturally acquired microchimerism: implications for transplantation outcome and novel methodologies for detection. *Chimerism*, *5*(2), pp. 24-39.

El-Hazmi, M. A., 2004. Ethics of genetic counseling—basic concepts and relevance to Islamic communities. *Annals of Saudi Medicine*, *24*(2), pp. 84-92.

Eyal, O., Berkenstadt, M., Reznik-Wolf, H., Poran, H., Ziv-Baran, T., Greenbaum, L., Yonath, H. and Pras, E., 2019. Prenatal diagnosis for de novo mutations: Experience from a tertiary center over a 10-year period. *Molecular genetics & genomic medicine*, *7*(4), p. e00573.

Falik-Zaccai, T. C., Kfir, N., Frenkel, P., Cohen, C., Tanus, M., Mandel, H., Shihab, S., Morkos, S., Aaref, S., Summar, M. L. and Khayat, M., 2008. Population screening in a Druze community: the challenge and the reward. *Genetics in Medicine*, *10*(12), p. 903.

Fan, H. and Chu, J. Y., 2007. A brief review of short tandem repeat mutation. *Genomics, Proteomics & Bioinformatics*, *5*(1), pp. 7-14.

Fokkema, I. F., Taschner, P. E., Schaafsma, G. C., Celli, J., Laros, J. F. and den Dunnen, J. T., 2011. LOVD v. 2.0: the next generation in gene variant databases. *Human mutation*, *32*(5), pp. 557-563. Fraser, F. C., 1974. Genetic counseling. *American journal of human genetics*, *26*(5), p. 636.

Forrester, M. B. and Merz, R. D., 2006. Use of prenatal diagnostic procedures in pregnancies affected with birth defects, Hawaii, 1986–2002. *Birth Defects Research Part A: Clinical and Molecular Teratology*, *76*(11), pp. 778-780.

Forrester, M. B. and Merz, R. D., 2007. Genetic counseling utilization by families with offspring affected by birth defects, Hawaii, 1986–2003. *American Journal of Medical Genetics Part A*, *143*(10), pp. 1045-1052.

Gesser-Edelsburg, A. and Shahbari, N. A. E., 2017. Decision-making on terminating pregnancy for Muslim Arab women pregnant with fetuses with congenital anomalies: maternal affect and doctor-patient communication. *Reproductive health*, 14(1), p. 49.

Gettings, K. B., Aponte, R. A., Vallone, P. M. and Butler, J. M., 2015. STR allele sequence variation: current knowledge and future issues. *Forensic Science International: Genetics*, *18*, pp. 118-130.

Glynn, A., Collins, V. and Halliday, J., 2009. Utilization of genetic counseling after diagnosis of a birth defect—trends over time and variables associated with utilization. *Genetics in Medicine*, *11*(4), p. 287.

Glynn, A., Saya, S. and Halliday, J., 2012. Use and non-use of genetic counseling after diagnosis of a birth defect. *American Journal of Medical Genetics Part A*, *158*(3), pp. 559-566.

Goodman, R. M., 1979. Genetic disorders among the Jewish people. Johns Hopkins University Press, Baltimore.

Goodman, R. M., 1984. Problems in medical genetic services as viewed from Israel. *Public Health Reports*, *99*(5), p. 460.

Grabber TM, Vanarsdall V. *Orthodontics Current Principles and Techniques. 4th ed.* St. Louis, U.S.A: Elsevier Publishers; 2005. Genetics; pp. 101–2.

Gross, R., Rosen, B. and Shirom, A., 2001. Reforming the Israeli health system: findings of a 3-year evaluation. *Health Policy*, *56*(1), pp. 1-20.

Hanany, M., Allon, G., Kimchi, A., Blumenfeld, A., Newman, H., Pras, E., Wormser, O., Birk, O. S., Gradstein, L., Banin, E. and Ben-Yosef, T., 2018. Carrier frequency analysis of mutations causing autosomal-recessive-inherited retinal diseases in the Israeli population. *European Journal of Human Genetics*, *26*(8), p. 1159.

Harper PS (2004). *Practical genetic counseling*, 6th ed. London, Arnold.

Hashiloni-Dolev, Y. and Weiner, N., 2008. New reproductive technologies, genetic counseling and the standing of the fetus: views from Germany and Israel. *Sociology of health & illness*, *30*(7), pp. 1055-1069.

Hong, K. W. and Oh, B. S., 2010. Overview of personalized medicine in the disease genomic era. *BMB reports*, *43*(10), pp. 643-648.

Huang, C., Jiang, W., Zhu, Y., Li, H., Lu, J., Yang, J. and Chen, Z., J., 2019. Pregnancy outcomes of reciprocal translocation carriers with two or more unfavorable pregnancy histories: before and after preimplantation genetic testing. *Journal of Assisted Reproduction and Genetics*. doi: 10.1007/s10815-019-01585-9.

HYUN, J., 2019. Doctors Discussing "the Root of Koreans": Medical Genetics and the Korean Origin, 1975-1987. *Korean Journal of Medical History*, *28*(2), pp. 551-590.

International Human Genome Sequencing Consortium, 2001. Initial sequencing and analysis of the human genome. *Nature*, *409*(6822), p. 860.

Isabella, M., Teresa, B. R., Silvano, B., Serena, B., Francesco, B., Stefania, B., Claudio, C., Maurizio, C., Elena, D. G., Marilena, p. and Licia, T., 2003. Utilization of genetic counseling by parents of a child or fetus with congenital malformation in North-East Italy. *American Journal of Medical Genetics Part A*, *121*(3), pp. 214-218.

Jaber, L., Dolfin, T., Shohat, T., Halpern, G. J., Reish, O. and Fejgin, M., 2000. Prenatal diagnosis for detecting congenital malformations: acceptance among Israeli Arab women. *The Israel Medical Association Journal: IMAJ*, *2*(5), pp. 346-350.

Jastaniah, W., 2011. Epidemiology of sickle cell disease in Saudi Arabia. *Annals of Saudi medicine*, *31*(3), pp. 289-293.

Kelly TE (1986). *Clinical genetics and genetic counseling*, 2nd ed. Chicago, Year Book Publishers.

Kessler, S., 1997. Psychological aspects of genetic counseling. XI. Nondirectiveness revisited. *American journal of medical genetics*, 72(2), pp. 164-171.

Khemiri, L., Larsson, H., Kuja-Halkola, R., D'Onofrio, B. M., Lichtenstein, P., Jayaram-Lindström, N. and Latvala, A., 2019. Association of Parental Substance Use Disorder with Offspring Cognition: A Population Family-based Study. *Addiction*.

Kinchen, K. S., Cooper, L. A., Levine, D., Wang, N. Y. and Powe, N. R., 2004. Referral of patients to specialists: factors affecting choice of specialist by primary care physicians. *The Annals of Family Medicine*, 2(3), pp. 245-252.

Landrum, M. J., Lee, J. M., Benson, M., Brown, G., Chao, C., Chitipiralla, S., Gu, B., Hart, J., Hoffman, D., Hoover, J. and Jang, W., 2015. ClinVar: public archive of interpretations of clinically relevant variants. *Nucleic acids research*, 44(D1), pp. D862-D868.

Lek, M., Karczewski, K. J., Minikel, E. V., Samocha, K. E., Banks, E., Fennell, T., O'Donnell-Luria, A. H., Ware, J. S., Hill, A. J., Cummings, B. B. and Tukiainen, T., 2016. Analysis of protein-coding genetic variation in 60,706 humans. *Nature*, 536(7616), p. 285.

Link, B. G. and Phelan, J., 1995. Social conditions as fundamental causes of disease. *Journal of health and social behavior*, pp. 80-94.

Marmot, M., 2005. Social determinants of health inequalities. *The Lancet*, 365(9464), pp. 1099-1104.

*Ministry of Health, Israel, 2019*, available from this website: https://www.health.gov.il/Subjects/Genetics/checks/Pages/genetic_counseling_instit ute.aspx.

Mirkin, S. M., 2007. Expandable DNA repeats and human disease. *Nature*, 447(7147), p. 932.

Muhsen, K., Na'amnah, W., Lesser, Y., Volovik, I., Cohen, D. and Shohat, T., 2010. Determinates of underutilization of amniocentesis among Israeli Arab women. *Prenatal Diagnosis: Published in Affiliation With the International Society for Prenatal Diagnosis*, 30(2), pp. 138-143.

Na'amnih, W., Muhsen, K., Tarabeia, J., Saabneh, A. and Green, M. S., 2010. Trends in the gap in life expectancy between Arabs and Jews in

Israel between 1975 and 2004. *International journal of epidemiology*, *39*(5), pp. 1324-1332.

National Insurance Institute, Israel (2015). *The 2014 poverty report*. Jerusalem: Israel National Insurance Institute.

New York-Mid-Atlantic Consortium for Genetic and Newborn Screening Services, 2009. *Understanding genetics: a New York, mid-Atlantic guide for patients and health professionals*. Lulu.com.

Othman, S. A., AlOjan, A., AlShammari, M. and Ammar, A., 2019. Awareness of spina bifida among family of. *Saudi Med J*, *40*(7), pp. 727-731.

Panter-Brick C (1991) Parental responses to consanguinity and genetic disease in Saudi Arabia. Soc Sci Med 33(11):1295–1302.

Phillips SE. Encyclopaedia of Sciences. USA: Eastern Virgenia Medical school, Norfolk, Virgenia; 2001. Genetic counseling; pp. 1–6.

Raz, A. E. and Atar, M., 2004. Cousin marriage and premarital carrier matching in a Bedouin community in Israel: attitudes, service development and educational intervention. *BMJ Sexual & Reproductive Health*, *30*(1), pp. 49-51.

Reed, S. C., 1974. A short history of genetic counseling. *Social Biology*, *21*(4), pp. 332-339.

Reed, S. C., 1979. A short history of human genetics in the USA. *American journal of medical genetics*, *3*(3), pp. 282-295.

Resta, R., Biesecker, B. B., Bennett, R. L., Blum, S., Estabrooks Hahn, S., Strecker, M. N. and Williams, J. L., 2006. A new definition of genetic counseling: National Society of Genetic Counselors' task force report. *Journal of genetic counseling*, *15*(2), pp. 77-83.

Resta, R. G., 2006, November. Defining and redefining the scope and goals of genetic counseling. In *American Journal of Medical Genetics Part C: Seminars in Medical Genetics* (Vol. 142, No. 4, pp. 269-275). Hoboken: Wiley Subscription Services, Inc., A Wiley Company.

Sa'd, J. A., Indelman, M., Pfendner, E., Falik-Zaccai, T. C., Mizrachi-Koren, M., Shalev, S., Amitai, D .B., Raas-Rothshild, A., Adir-Shani, A., Borochowitz, Z. U. and Gershoni-Baruch, R., 2006. Molecular epidemiology of hereditary epidermolysis bullosa in a Middle Eastern population. *Journal of investigative dermatology*, *126*(4), pp. 777-781.

Sagar, A. D. and Najam, A., 1998. The human development index: a critical review1. *Ecological economics*, *25*(3), pp. 249-264.

Schoeffel, K., Veach, p. M., Rubin, K. and LeRoy, B., 2018. Managing couple conflict during prenatal counseling sessions: an investigation of genetic counselor experiences and perceptions. *Journal of genetic counseling*, *27*(5), pp. 1275-1290.

Seidahmed, M. Z., Abdelbasit, O. B., Shaheed, M. M., Alhussein, K. A., Miqdad, A. M., Khalil, M. I., Al-Enazy, N. M. and Salih, M. A., 2014. Epidemiology of neural tube defects. *Saudi medical journal*, *35*(Suppl 1), p. S29.

Shalev, S. A., 2019. Characteristics of genetic diseases in consanguineous populations in the genomic era: Lessons from Arab communities in North Israel. *Clinical genetics*, *95*(1), pp. 3-9.

Sharkia, R., Azem, A., Kaiyal, Q., Zelnik, N. and Mahajnah, M., 2010. Mental retardation and consanguinity in a selected region of the Israeli Arab community. *Central European journal of medicine*, *5*(1), pp. 91-96.

Sharkia, R., Mahajnah, M., Athamny, E., Khatib, M., Sheikh-Muhammad, A. and Zalan, A., 2016. Changes in marriage patterns among the Arab community in Israel over a 60-year period. *Journal of biosocial science*, *48*(2), pp. 283-287.

Sharkia, R., Mahajnah, M., Sheikh-Muhammad, A., Khatib, M. and Zalan, A., 2018. Trends in the prevalence of type 2 diabetes mellitus among Arabs in Israel: A community health survey. *Glob Adv Res J Med Med Sci*, *7*(4), pp. 98-104.

Sharkia, R., Mahajnah, M., Zalan, A., Athamna, M., Azem, A., Badarneh, K. and Faris, F., 2013. Comparative screening of FMF mutations in various communities of the Israeli society. *European journal of medical genetics*, *56*(7), pp. 351-355.

Sharkia, R., Mahajnah, M., Zalan, A., Sourlis, C., Bauer, p. and Schöls, L., 2014. Sanfilippo type A: new clinical manifestations and neuroimaging findings in patients from the same family in Israel: a case report. *Journal of medical case reports*, *8*(1), p. 78.

Sharkia, R., Shalev, S. A., Zalan, A., Watemberg, N., Urquhart, J. E., Daly, S. B., Bhaskar, S. S., Williams, S. G., Newman, W. G., Spiegel, R. and Azem, A., 2017. Homozygous mutation in *PTRH2* gene causes progressive sensorineural deafness and peripheral neuropathy. *American Journal of Medical Genetics Part A*, *173*(4), pp. 1051-1055.

Sharkia, R., Tarabeia, J., Zalan, A., Atamany, E., Athamna, M. and Allon-Shalev, S., 2015. Factors affecting the utilization of genetic counseling services among Israeli Arab women. *Prenatal diagnosis*, *35*(4), pp. 370-375.

Sharkia, R., Zaid, M., Athamna, A., Cohen, D., Azem, A. and Zalan, A., 2008. The changing pattern of consanguinity in a selected region of the Israeli Arab community. *American Journal of Human Biology*, *20*(1), pp. 72-77.

Sharkia, R., Zalan, A., Jabareen-Masri, A., Hengel, H., Schöls, L., Kessel, A., Azem, A. and Mahajnah, M., 2019. A novel biallelic loss-of-function mutation in *TMCO1* gene confirming and expanding the phenotype spectrum of cerebro-facio-thoracic dysplasia. *American Journal of Medical Genetics Part A*. doi: 10.1002/ajmg.a. 61168.

Sharkia, R., Zalan, A., Jabareen-Masri, A., Zahalka, H. and Mahajnah, M., 2018. A new case confirming and expanding the phenotype spectrum of ADAT3-related intellectual disability syndrome. *European journal of medical genetics*. https://doi.org/10.1016/j.ejmg.2018.10.001.

Sharony, R., Engel, O., Litz-Philipsborn, S., Sukenik-Halevy, R., Biron-Shental, T. and Evans, M. I., 2018. The impact of third-trimester genetic counseling. *Archives of gynecology and obstetrics*, *297*(3), pp. 659-665.

Sheiner, E., Shoham-Vardi, I., Weitzman, D., Gohar, J. and Carmi, R., 1998. Decisions regarding pregnancy termination among Bedouin couples referred to third level ultrasound clinic. *European Journal of Obstetrics & Gynecology and Reproductive Biology*, *76*(2), pp. 141-146.

Shiels, A. and Hejtmancik, J. F., 2019. Biology of Inherited Cataracts and Opportunities for Treatment. *Annual Review of Vision Science.* 15;5:123-149.

Siani, M. and Assaraf, O. B. Z., 2017. A qualitative look into Israeli genetic experts' insights regarding culturally competent genetic counseling and recommendations for its enhancement. *Journal of genetic counseling*, 26(6), pp. 1254-1269.

Simpson, J. L., Rechitsky, S. and Kuliev, A., 2019. Before the beginning: the genetic risk of a couple aiming to conceive. *Fertility and Sterility*, 112(4), pp. 622-630.

Sirdah, M. M., 2014. Consanguinity profile in the Gaza Strip of Palestine: large-scale community-based study. *European journal of medical genetics*, 57(2-3), pp. 90-94.

Sutton HE. An Introduction to Human Genetics. 4th ed. San Diego: Harcourt Brace Jovanovich; 1998. pp. 23–56.

Tadmouri, G. O., Nair, P., Obeid, T., Al Ali, M. T., Al Khaja, N. and Hamamy, H. A., 2009. Consanguinity and reproductive health among Arabs. *Reproductive health*, 6(1), p. 17.

Tarabeia, J., Amitai, Y., Green, M., Halpern, G. J., Blau, S., Ifrah, A., Rotem, N. and Jaber, L., 2004. Differences in infant mortality rates between Jews and Arabs in Israel, 1975-2000. *IMAJ-RAMAT GAN-.*, 6(7), pp. 403-407.

Teebi, A. S. and Teebi, S. A., 2005. Genetic diversity among the Arabs. *Public Health Genomics*, 8(1), pp. 21-26.

Teebi, A. S. ed., 2010. *Genetic disorders among Arab populations.* Springer Science & Business Media.

Treister-Goltzman, Y. and Peleg, R., 2014. Health and morbidity among Bedouin women in southern Israel: a descriptive literature review of the past two decades. *Journal of community health*, 39(4), pp. 819-825.

Treister-Goltzman, Y. and Peleg, R., 2015. Literature review of type 2 diabetes mellitus among minority Muslim populations in Israel. *World journal of diabetes*, 6(1), p. 192.

United Nations Educational Scientific and Cultural Organization (1999). *Universal declaration on the human genome and human rights.* UNESCO (A/RES/53/152).

United Nations Educational Scientific and Cultural Organization (2003). *International declaration of human genetic data.* Paris, UNESCO.

Vardi-Saliternik, R., Friedlander, Y. and Cohen, T., 2002. Consanguinity in a population sample of Israeli muslim Arabs, christian Arabs and druze. *Annals of human biology, 29*(4), pp. 422-431.

Willems, T., Gymrek, M., Highnam, G., Mittelman, D., Erlich, Y. and 1000 Genomes Project Consortium, 2014. The landscape of human STR variation. *Genome research, 24*(11), pp. 1894-1904.

World Health Organization (1996). *Control of hereditary diseases.* Geneva, World Health Organization (Technical Report Series, No. 865).

World Health Organization (2002). *Genomics and world health.* Geneva, World Health Organization.

World Health Organization Scientific Group on the Control of Hereditary Diseases (1993). *Control of hereditary diseases: report of a WHO scientific group.* Geneva, World Health Organization.

World Health Organization, 1962. *The teaching of genetics in the undergraduate medical curriculum and in postgraduate training: first report of the Expert Committee on Human Genetics* (meeting held in Geneva from 29 November to 4 December 1961). World Health Organization.

Yitzhaki S. Israel in Figures: 2010. *Jerusalem: Central Bureau of Statistics*, 2010.

Yitzhaki, S., 1983. On an extension of the Gini inequality index. *International economic review*, pp. 617-628.

Zlotogora, J., 2002. Parental decisions to abort or continue a pregnancy with an abnormal finding after an invasive prenatal test. *Prenatal Diagnosis: Published in Affiliation With the International Society for Prenatal Diagnosis, 22*(12), pp. 1102-1106.

Zlotogora, J., Barges, S., Bisharat, B. and Shalev, S. A., 2006. Genetic disorders among Palestinian Arabs. 4: Genetic clinics in the community. *American Journal of Medical Genetics Part A*, *140*(15), pp. 1644-1646.

Zlotogora, J., Hujerat, Y., Barges, S., Shalev, S. A. and Chakravarti, A., 2007. The fate of 12 recessive mutations in a single village. *Annals of human genetics*, *71*(2), pp. 202-208.

Zuckerman, S., Lahad, A., Shmueli, A., Zimran, A., Peleg, L., Orr-Urtreger, A., Levy-Lahad, E. and Sagi, M., 2007. Carrier screening for Gaucher disease: lessons for low-penetrance, treatable diseases. *JAMA*, *298*(11), pp. 1281-1290.

# INDEX

## Related Nova Publications

RESISTIN: STRUCTURE, FUNCTION AND ROLE IN DISEASE

**EDITOR:** Gerald Maldonado

**SERIES:** Biochemistry Research Trends

**BOOK DESCRIPTION:** *Resistin: Structure, Function and Role in Disease* opens by discussing resistin's therapeutic properties. Because evidence appears to suggest that resistin is a proinflammatory cytokine, resistin may impact metabolic disease.

**SOFTCOVER ISBN:** 978-1-53614-543-4
**RETAIL PRICE:** $95

AN ESSENTIAL GUIDE TO CYTOGENETICS

**EDITORS:** Naomi Norris and Carmen Miller

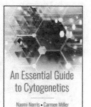

**SERIES:** Genetics – Research and Issues

**BOOK DESCRIPTION:** *An Essential Guide to Cytogenetics* explores the use of cytogenetic data for studies of frogs as well as the insights that hypotheses of phylogenetic relationships have added to this issue.

**SOFTCOVER ISBN:** 978-1-53613-370-7
**RETAIL PRICE:** $95

*To see a complete list of Nova publications, please visit our website at www.novapublishers.com*

## Related Nova Publications

### DNA: BACKGROUND, LAWS AND BACKLOG OF EVIDENCE

**EDITOR:** Tomáš Koláček

**SERIES:** Genetics – Research and Issues

**BOOK DESCRIPTION:** Deoxyribonucleic acid, or DNA, is the fundamental building block for an individual's entire genetic makeup. DNA is a powerful tool for law enforcement investigations because each person's DNA is different from that of every other individual (except for identical twins).

**HARDCOVER ISBN:** 978-1-53616-117-5
**RETAIL PRICE:** $160

### ENCYCLOPEDIA OF GENETICS: NEW RESEARCH (8 VOLUME SET)

**EDITOR:** Heidi Carlson

**SERIES:** Genetics – Research and Issues

**BOOK DESCRIPTION:** This 8 volume encyclopedia set presents important research on genetics. Some of the topics discussed herein include the speciation of Arabian gazelles, tau alternative splicing in Alzheimer's disease, Cornelia de Lange syndrome and autosomal dominant polycystic kidney disease.

**HARDCOVER ISBN:** 978-1-53614-451-2
**RETAIL PRICE:** $1,380

*To see a complete list of Nova publications, please visit our website at www.novapublishers.com*